PETROLEUM ECONOMICS AND RISK ANALYSIS

To John

Thank you so much for all your help in pulling this together — a long time coming!

Very best regards

Mark.

DEVELOPMENTS IN PETROLEUM SCIENCE 71

PETROLEUM ECONOMICS AND RISK ANALYSIS

A Practical Guide to E&P Investment Decision-Making

MARK COOK
SPE (Society of Petroleum Engineers),
Distinguished Lecturer
TRACS International Ltd., Aberdeen, UK
Delta-T Energy Consultancy Ltd., Dublin, Ireland

ELSEVIER

Elsevier
Radarweg 29, PO Box 211, 1000 AE Amsterdam, Netherlands
The Boulevard, Langford Lane, Kidlington, Oxford OX5 1GB, United Kingdom
50 Hampshire Street, 5th Floor, Cambridge, MA 02139, United States

Copyright © 2021 Elsevier B.V. All rights reserved.

No part of this publication may be reproduced or transmitted in any form or by any means, electronic or mechanical, including photocopying, recording, or any information storage and retrieval system, without permission in writing from the Publisher. Details on how to seek permission, further information about the Publisher's permissions policies and our arrangements with organizations such as the Copyright Clearance Center and the Copyright Licensing Agency, can be found at our website: www.elsevier.com/permissions.

This book and the individual contributions contained in it are protected under copyright by the Publisher (other than as may be noted herein).

Notices
Knowledge and best practice in this field are constantly changing. As new research and experience broaden our understanding, changes in research methods, professional practices, or medical treatment may become necessary.

Practitioners and researchers must always rely on their own experience and knowledge in evaluating and using any information, methods, compounds, or experiments described herein. In using such information or methods they should be mindful of their own safety and the safety of others, including parties for whom they have a professional responsibility.

To the fullest extent of the law, neither the Publisher, nor the authors, contributors, or editors, assume any liability for any injury and/or damage to persons or property as a matter of products liability, negligence or otherwise, or from any use or operation of any methods, products, instructions, or ideas contained in the material herein.

Library of Congress Cataloging-in-Publication Data
A catalog record for this book is available from the Library of Congress

British Library Cataloguing-in-Publication Data
A catalogue record for this book is available from the British Library

ISBN: 978-0-12-821190-8
ISSN: 0376-7361

For information on all Elsevier publications visit our website at
https://www.elsevier.com/books-and-journals

Publisher: Candice Janco
Acquisitions Editor: Amy Shapiro
Editorial Project Manager: Grace Lander
Production Project Manager: Sruthi Satheesh
Cover Designer: Matthew Limbert

Typeset by TNQ Technologies

Contents

Acknowledgement ix

1. Introduction — 1
1.1 Structure of the book — 1
1.2 Gaining access to E&P opportunity — 4
1.3 Exploration — 5
1.4 Appraisal planning — 6
1.5 Evaluating field development opportunities — project cash flow — 7
1.6 Risk analysis — 10
1.7 Evaluating incremental project opportunities — 12
1.8 Decommissioning — 12
1.9 Portfolio investment — 13
1.10 Abbreviations and units — 14
Reference — 14

2. Trends in global energy supply and demand — 15
2.1 Supply and demand curves for general commodities — 15
2.2 Historical trends in global energy supply and demand — 20
2.3 Forecasts of global energy supply and demand — 33
References — 42

3. Gaining access to E&P opportunity — 43
3.1 Options for accessing E&P investment opportunity — 43
3.2 Strategic choices for investment — 44
3.3 Bidding processes — 45
3.4 Acquisition and disposal — 51
3.5 Merger — 55
3.6 Joint venture and partnership — 57
3.7 Unitisation and equity determination — 58
3.8 Financing the project — 61
References — 67

4. Exploration — 69
4.1 Introduction — 69
4.2 Risk and uncertainty in exploration — 70
References — 100

v

5. Appraisal planning — 101
5.1 Reservoir appraisal and project assessment in context — 101
5.2 Justifying reservoir appraisal assuming perfect information — 103
5.3 Incorporating imperfect information — 114
5.4 Reservoir appraisal planning — 122
References — 131

6. Project cash flow — 133
6.1 Cash flow components — 133
6.2 Production profile and sales price — the revenue source — 136
6.3 Capex and opex — the technical costs — 145
6.4 Fiscal systems — the host government take — 153
6.5 Constructing the project cash flow — 176
6.6 Discounting — the time value of money — 187
6.7 Incorporating inflation into the project cash flow — 196
References — 204

7. Economic indicators from the DCF — 207
7.1 Indicators from the cash flow and discounted cash flow — 207
7.2 Indicators of efficiency — 208
7.3 Summary of economic indicators — 221
7.4 Incorporating decommissioning cost — 224
7.5 Choosing between alternative projects — 225
7.6 Distinguishing net cash flow from net income — 228
References — 229

8. Risk analysis and decision-making for development — 231
8.1 Development planning using a scenario approach — 231
8.2 Describing subsurface uncertainty — reservoir realisations — 233
8.3 Creating development concepts — 240
8.4 Applying investment themes — 242
8.5 Selecting the favoured concept — 244
8.6 Risk management of the development concept — 245
References — 274

9. Incremental projects and decommissioning — 275
9.1 Identifying incremental projects — 275
9.2 Incremental project economics — 281
9.3 Ranking incremental project opportunities — 287
9.4 Decommissioning — 295
References — 306

10. Portfolio management — **309**

- 10.1 Managing value and risk — 309
- 10.2 Portfolio effect on volumes and risk — 310
- 10.3 Portfolio effect on value and risk — 312
- 10.4 The efficient frontier — balancing the portfolio to match strategy — 324
- 10.5 Gambler's ruin and exploration portfolios — 330
- References — 335

Abbreviations, symbols and units — *337*
Index — *341*

Acknowledgement

This book represents some forty years of my learning from asset evaluation studies and teaching in the upstream oil and gas industry. I have been privileged to work alongside the highest calibre professionals in TRACS and client companies, and to be involved in projects ranging from exploration to late-life field management. This book aims to capture techniques for economic evaluation and risk analysis practised across the industry.

I had already written much of the technical content into our training manuals, and have finally incorporated the material into this publication. As in the classroom, I believe that illustrations and examples are a key to effective learning, and I am most grateful to Susan McLafferty, Fiona Swapp and Anne Cook for their graphic design input, and to TRACS, BP, IEA and others for granting permission to use a selection of their images. I owe a huge debt of gratitude to Dr John Gallivan for his tireless work in reviewing every detail of this book. Finally, I thank my training course participants whose stimulating feedback and experience has fed into the development of this book.

CHAPTER 1

Introduction

Contents

1.1 Structure of the book	1
1.2 Gaining access to E&P opportunity	4
1.3 Exploration	5
1.4 Appraisal planning	6
1.5 Evaluating field development opportunities – project cash flow	7
1.6 Risk analysis	10
1.7 Evaluating incremental project opportunities	12
1.8 Decommissioning	12
1.9 Portfolio investment	13
1.10 Abbreviations and units	14
Reference	14

1.1 Structure of the book

When we wrote *Hydrocarbon Exploration and Production* (Jahn, Cook, Graham, Ref. [1]), it was designed to introduce the technical aspects of the upstream oil and gas business, with a chapter dedicated to the commercial evaluation of projects. This book focuses on project evaluation using the techniques of petroleum economics and methods of identifying and managing both technical and commercial risks associated with exploration and production (E&P) investments.

The two books have a common theme running through the chapters, in that they follow the field lifecycle, from setting up the E&P opportunity for the investor, through the stages of exploration, appraisal, field development, production and finally, decommissioning.

The field lifecycle can be represented in terms of timing and investment as shown in Fig. 1.1. At each stage of the lifecycle, methods of project evaluation and risk management are described, with worked examples. While generally focussing on a single prospect or field development the book also demonstrates the benefits of creating a portfolio of assets to reduce overall risk.

The axis scales in Fig. 1.1 represent a midsized E&P investment. Two features differentiate the nature of the E&P business: the duration of the

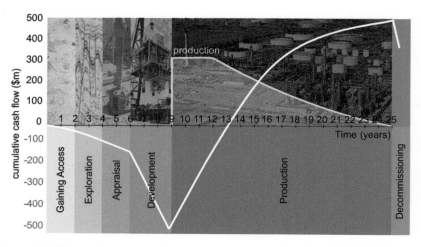

Figure 1.1 The exploration and production (E&P) field lifecycle. *(Adapted from Hydrocarbon Exploration and Production, Elsevier, 2008.)*

field lifecycle, and the magnitude of the investment. The y-axis represents the cumulative cash flow of the project, showing that the investor is exposed to some $500 million prior to receiving any revenue from production. In this example, the exploration and appraisal activity leads to a field development. Many exploration activities, however, do not deliver any reward. The time axis shows the lifetime of the project running to 25 years, but total investment is only recovered after 13 years.

Large fields can run to billions of dollars of exposure and run for many decades. The Johan Svedrup Field was discovered in the Norwegian sector of the North Sea in 2010. Operated by Equinor, it started production January 2020, with an estimated production lifetime of 50 years. Maximum production is expected to be 660 thousand barrels per day (Mb/d), with an ultimate recovery estimated at 2.7 billion barrels (Bbbl). The first phase of investment, including four platforms, wells and export systems, will cost around $9 billion [2].

The scale of investment and the time to recover the initial investment demonstrate that the E&P business is both capital intensive and risky. It is for these reasons that investors analyse opportunities with care and consistency, using methods described in this book. The techniques are applicable to the range of investors, covering small independent enterprises, midsized companies, international oil companies (IOCs) and national oil companies (NOCs).

To provide an overview of the E&P business and context for further discussion, Chapter 2 describes global oil and gas supply and demand trends up to 2020. It summarises forecasts, made by professional groups, of the future of fossil fuels as the world enters a period of transition in energy supply in pursuit of the Paris Agreement on Climate Change. While oil and gas will compete with renewable sources of energy, they continue to have a strong role to play in supplying the world's increasing energy demand as population grows and lesser-developed countries industrialise and move to urban dwelling.

The book is organised as shown in Table 1.1. The reader may choose to focus on a chapter that pertains to the project they are considering.

Table 1.1 Layout of the book.

Chapter	Title	Content
1	Introduction	Purpose and layout of the book.
2	Trends in global energy supply and demand	Summary of historical global energy supply and demand and prediction of future trends.
3	Gaining access to E&P opportunity	Setting up the conditions to be able to bid on exploration acreage or participate in development projects. Methods of finance available in the exploration and production (E&P) sector.
4	Exploration	Balancing risk and reward, as exploration expenditure may result in discovery or failure. Calculating expected monetary value (EMV).
5	Appraisal planning	Determining how to reduce reservoir and commercial uncertainties in a cost-effective manner, prior to committing to a development plan. Applying value of information (VOI) techniques.
6	Project cash flow	The fundamentals of calculating a net present value (NPV) from a discounted cash flow (DCF) for a development project, incorporating revenues and all technical and fiscal costs and revenues.
7	Economic indicators from the DCF	Value and efficiency measures for a project net cash flow.
8	Risk analysis and decision-making for development	Contexts for the definition of risk. A structured approach to identifying and managing technical and commercial risks.
9	Incremental projects and decommissioning	DCF techniques applied to incremental projects, which are additional to the initial major investment. Estimating and incorporating decommissioning costs, while meeting regulatory requirements.
10	Portfolio management	How building a portfolio of assets reduces risk.

Because the discounted cash flow (DCF) approach used in petroleum economics is fundamental, its derivation and importance are briefly summarised in this introduction. The details of the DCF technique are provided in Chapter 6.

1.2 Gaining access to E&P opportunity

Access to E&P investment opportunity can be gained at any stage of the oil or gas field lifecycle, through direct acquisition of a project or a company, or by taking a share of a project as a partner. A clear company strategy drives the choice of asset type, geography, geology and level of participation to target when attempting to gain access to the E&P business.

The first potential point of entry as an active participant is at the exploration phase, through bidding in licencing rounds, announced periodically by host nations. The bid is based on the risked value of the licence area or concession on offer, for which an expected monetary value (EMV) should be estimated. In gaining entry to field development projects or projects with existing production, the basis of an offer becomes the net present value (NPV).

Access may be gained by acquiring an existing company, whose valuation is the sum of the portfolio it owns. The value of an E&P company's portfolio is dominated by the NPV of its projects, based on the production of remaining reserves. Reserves carry a range of uncertainty and projects have different classes of maturity. Chapter 3 describes reserves classification using a widely accepted methodology. Reserves are converted into value using DCF methods detailed in Chapter 6, but summarised briefly in this introductory chapter (Section 1.5).

The industry has a strong history of mergers between companies as a method of growth. Mergers are intended to add value through increased efficiency and access to wider opportunity. Forming joint ventures (JVs) allows companies to share capital investment and expertise, and JVs between independent oil companies and NOCs have been widespread in recent decades.

Converting every E&P opportunity requires finance, and Chapter 3 introduces various forms of finance available to the E&P sector, identifying their suitability to different types of investor. In addition to traditional forms of finance, such as loans (debt) and equity (shares), the finance industry has developed a number of bespoke methods. These include mezzanine financing, reserve-based lending and volumetric production payment.

1.3 Exploration

Governments or their representatives periodically announce bidding rounds for licence blocks or concessions, inviting companies to compete to gain access to E&P opportunity through exploration. Bids are competitive, and the bidders aim is to win the bid, without overbidding. The technical and commercial evaluation of the block determines the potential reward, which would be the value derived from developing an oil or gas field. The risk, however, is that no commercial discovery is made, and the exploration activity yields no returns. Chapter 4 describes how the bidder makes the risk-reward calculation, and takes into account the level of competition for the block in formulating their bid.

In the context of exploration, the risk is the chance of discovering hydrocarbons, known as the probability of geological success (P_g). The probability of making a discovery can be as low as 1 in 10 in an unexplored region, and as high as 1 in 2 in a mature region. The uncertainty relates to the volume of recoverable hydrocarbons, if a discovery is made. This can easily vary by a factor of 10 between a low and high case estimate, at the exploration stage. Chapter 4 summarises how geoscientists and engineers estimate P_g and the range of volumes of recoverable hydrocarbons.

The monetary reward from a successful exploration venture would be a net cash flow from a field development. This reward can be represented by a NPV, using the DCF techniques detailed in Chapter 6. The chance of making a discovery is combined with the value of a development by multiplying the NPV by the probability of making the discovery (P_g), to yield the EMV of the opportunity. The EMV is the risked value of the opportunity, which will form the basis of the bid.

In this approach, the EMV does not include the cost of the bid, which will usually include a commitment to shoot a seismic survey of the block and drill a number of commitment wells, and in many cases a promise of a signature bonus payable to the government in case of block award. If the cost of the bid were equal to the EMV, then on average the bidder will break even. Therefore, the bidder will wish to make a bid whose cost is some fraction of the EMV. The fraction will depend on the perceived level of competition and the enthusiasm that the bidder has for the block, which is often a matter of taste. Chapter 4 describes a structured methodology using decision tree analysis for combining geological risk, volumetric uncertainty and project value to determine EMV, which is the guide to formulating the bid.

1.4 Appraisal planning

Chapter 5 discusses appraisal in the context of reducing uncertainty in a discovered reservoir, but also explains another context in which the word appraisal is sometimes used, in Stage Gate processes.

Following an exploration venture that finds hydrocarbons, thus proving geological success, the next challenge is to describe the reservoir and fluid properties in more detail, to enable suitable development planning. Reservoir appraisal activities include shooting higher definition seismic, drilling appraisal wells, gathering fluid and core samples and well testing.

A key objective of reservoir appraisal planning is to reduce uncertainty in the reservoir description, which allows the design of the right-sized production facilities, thereby improving project value. The value is improved by avoiding the risk of over- or under-expenditure on facilities, or avoiding agreeing to sales contracts that do not match the potential of the field.

A simple approach to reservoir appraisal is to demonstrate that the field is sufficiently large and productive to support a development that will make a break-even economic case, the so-called minimum commercial field size. This provides some comfort in making a decision to develop the appraised field, but does not optimize the development plan.

Appraisal activity is both an expense and a delay to commencing project development. Its justification can be tested using the value of information (VOI) techniques introduced in Chapter 5. The maximum VOI of appraisal activity is the value of the development project which takes into account the appraisal information minus the value of the development project without the appraisal information. Again, decision tree analysis is a tool for presenting VOI calculations in a transparent manner. Considering and justifying appraisal information is not restricted to reservoir appraisal, it extends to improving understanding of commercial, HSE (health, safety and environment) and political uncertainties.

In some cases, the appraisal studies will conclude that a development cannot succeed commercially, and that the discovery will not progress along the field lifecycle of Fig. 1.1. While this may be a disappointment, the appraisal will add value by avoiding investing in a loss-making development.

It is often assumed that the information derived from appraisal activity is reliable, so-called perfect information. This is not necessarily the case, and the reliability of the appraisal information may significantly reduce its value. Chapter 5 provides examples of using Bayesian revision of probability to estimate posterior probabilities, which can incorporate an estimate of the reliability of imperfect information.

1.5 Evaluating field development opportunities — project cash flow

Chapter 6 details the components of project cash flow analysis, used to determine the attractiveness of investing in a field development. However, to assist the understanding of the terms NPV and EMV used in Chapters 4 and 5, this introduction will briefly summarise how the DCF technique is used to derive NPV and some other key economic indicators, using an example of an oil field, developed under a tax and royalty fiscal system. This example purposely presents a short project, for clarity.

In deciding whether to progress to development of an appraised field, only the point-forward elements of the project cash flow are considered. The incurred exploration and appraisal costs are considered to be sunk costs. This is a key principle of petroleum economics, only to look forward, never to look back.

Capital expenditure (capex) will be required to construct production facilities and drill wells prior to achieving the first revenue from oil production, which occurs in Year 3 in Fig. 1.2. Capex continues beyond the first oil production date to complete the drilling of development wells. A short production plateau is maintained through Years 4 and 5, and then production declines with time. The annual gross revenue is the annual oil production multiplied by the assumed oil price. On an annual basis:

$$\text{gross revenue}(\$) = \text{production}(\text{bbl}) \times \text{oil price}(\$/\text{bbl})$$

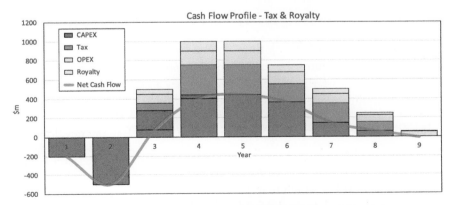

Figure 1.2 Cash flow forecast for an oil project with tax and royalty.

Once production commences in Year 3, operating expenditure (opex) is required to pay for facilities operations and maintenance, oil transportation or tariff costs, salaries, rental costs and other overheads. Annual opex is often based on a fixed cost, related to the value of plant, and a variable cost, related to the annual production. Total opex is assumed to decline as production declines, as it does in the example.

The fiscal system determines how the host government claims its share of the revenues from production, and the system depends upon the agreement between the oil field asset owners and the host government. Various fiscal systems are described in Section 6.4, but the example assumes simple tax and royalty, paid under a concessionary agreement. Details of Production Sharing Contracts (PSCs) are provided in Section 6.4.3.

Once production commences, the host government claims a percentage of the gross revenue stream. This is often in the range 10%–20%. On an annual basis:

$$\text{royalty}(\$) = \text{gross revenue}(\$) \times \text{royalty rate}(\%)$$

The government will also claim tax, based on taxable income. Certain tax deductions are allowable against gross revenue to calculate taxable income. These include the royalty already paid to the government, the opex incurred in running the project and an allowance for the capex invested in the project. The details of the typical capital allowance calculation are covered in Section 6.4.2. On an annual basis:

$$\text{tax}(\$) = \text{taxable income}(\$) \times \text{tax rate}(\%)$$
$$\text{taxable income}(\$) = \text{gross revenue} - \text{tax deductions}(\$)$$
$$\text{tax deductions}(\$) = \text{royalty} + \text{opex} + \text{capital allowance}(\$)$$

Finally, the annual net cash flow, shown by the green line in Fig. 1.2, is calculated by subtracting the capex, royalty, opex and tax from the gross revenue. The first year in which the project has a positive net cash flow is Year 3, sometimes called the pay-as-you-go time.

$$\text{net cash flow}(\$) = \text{gross revenue} - \text{capex} - \text{royalty} - \text{opex} - \text{tax}(\$)$$

In Year 9, the net cash flow becomes negative as the royalty and opex exceed the gross revenue. The project has gone beyond the economic limit (EL), which is at the end of Year 8, and the owners should consider decommissioning the field. The economic lifetime of the project is 8 years.

The annual net cash flow is presented on a cumulative basis in Fig. 1.3. This shows that by the end of the project, which should be curtailed at the end of Year 8, the cumulative net cash flow is $791 m, ignoring decommissioning cost. The investor is $791 m better off at the end of Year 8, thanks to the project. The cumulative net cash flow curve in green shows that the project recovers its initial cost at the end of Year 4, so the payback time, or payout time, is 4 years. The lowest value on the cumulative net cash flow curve is $700 m, known as the maximum exposure of the project, representing the amount of money the investor will require to carry out this project, and also the size of the loss if the project were to fail at the worst possible moment.

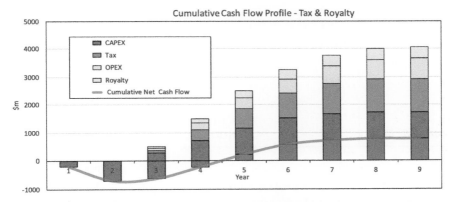

Figure 1.3 Cumulative cash flow forecast for an oil project with tax and royalty.

A key assumption in the analysis so far is that the dollars spent or received in each year have an equivalent value. As any investor knows, this is not true — we favour $100 now to $100 in Year 8, since there is an opportunity to invest the $100 now and allow compound interest to accrue over the interim 8 years and exceed the alternative offer of the $100 in Year 8. This is sometimes referred to as 'the time value of money'.

Recognising the disparity between dollars today and dollars in the future, the net cash flow is discounted, to calculate the Present Value (PV) of each annual net cash flow at the reference date of the project, which is the beginning of Year 1. A simple formula is applied to each future net cash flow (C_t) to calculate its PV (C_0) using a discount rate (r), where time (t) is the year in which the net cash flow occurs, counted from the reference date. The discount rate is the cost of capital to the company, which is fully explained in Section 6.6.

$$\text{Present Value } (C_0) = \frac{\text{future sum } (C_t)}{(1+r)^t} (\$)$$

Each year's PV is added together to yield the NPV of the project, as shown in Table 1.2, which assumes a discount rate of 8%, and net cash flows occurring at the end of each year.

Table 1.2 Net Present Value (NPV) calculation from a project net cash flow.

Time (year)	Net cash flow ($)	PV calculation ($)	Present value ($)
1	−200	$= -200/(1 + 0.08)^1$	−185.2
2	−500	$= -500/(1 + 0.08)^2$	−428.7
3	77	$= 77/(1 + 0.08)^3$	61.1
4	402	$= 402/(1 + 0.08)^4$	295.6
5	442	$= 442/(1 + 0.08)^5$	301.1
6	362	$= 362/(1 + 0.08)^6$	228.2
7	147	$= 147/(1 + 0.08)^7$	85.9
8	61	$= 61/(1 + 0.08)^8$	33.1
Net cash flow	791	NPV	391.1

NPV is a measure of project value, accounting for all elements of the project cash flow, and the time value of money. The inclusion of inflationary assumptions in costs and product prices is covered in Section 6.7. Chapter 7 details other important economic indicators, including efficiency measures such as Internal Rate of Return (IRR), Real Rate of Return (RROR), Unit Technical Cost (UTC) and capital efficiency ratios. The chapter summarises all petroleum economic indicators used in the analysis of a project cash flow, and shows how decommissioning cost is included in the cash flow calculation.

1.6 Risk analysis

The cumulative cash flow in Fig. 1.1 shows that the biggest investment during an oil or gas field lifecycle occurs during the development phase. Consequently, a realistic understanding of uncertainties and management of the associated risks is crucial to selecting a robust field development plan.

Even after effective reservoir appraisal, there will still be uncertainty in the understanding of the reservoir, and the forecasts of the production profile will have a significant range. This creates a challenge for the engineers who are designing process facilities and export systems, since there is a limit to the flexibility in flow rate that any one design can offer.

Chapter 8 recommends a scenario approach to field development planning, combining a set of plausible reservoir descriptions (called realisations) with alternative development concepts. A development concept describes a choice of reservoir management plan, well type, production facilities and export system. This approach creates a matrix of possibilities, each of which is evaluated using the petroleum economics tools. This practice is performed during the Select stage of the Stage Gate process described in Section 5.1, with the intention of selecting one development concept to progress to the Design stage for detailed design.

The final decision on which development concept to take through the Select Gate to the Define Stage is influenced by the investor's criteria for success. Chapter 8 describes a methodology for selecting a development concept that is robust to the range of reservoir uncertainty, and which meets the key investment criteria.

For a given development concept, spider diagrams are introduced to present how the project NPV is impacted by uncertainty in assumptions made in the economic model. Where the impact is significant, further risk management is necessary. In the context of development planning, risk is defined as the product of impact (usually measured in dollars) and probability of occurrence of the risk event (a number between 0 and 1.0). This forms the basis of Quantitative Risk Assessment (QRA). It is of course entirely different from the exploration geoscientist's definition of the word risk (P_g) used in Chapter 4.

The QRA methodology is used to design controls and contingencies that reduce the probability and the impact of any risk event that threatens the success of a project. The Bow Tie model provides a useful framework for this analysis. Chapter 8 provides examples of the use of the Bow Tie model, capturing risk management actions on a risk register, which is a valuable tracking tool. The risk levels before and after the risk management actions are presented on a risk matrix.

To reinforce the idea that the definition of risk depends on the context, the end of Chapter 8 introduces yet another definition of risk. This definition is often used by financial analysts, and is the standard deviation of value (σ) divided by the mean value (μ), known as the coefficient of variation (CV). This becomes useful in considering how a portfolio investment can reduce risk, discussed in Chapter 10.

1.7 Evaluating incremental project opportunities

Following the major investment in a project and the start of production, incremental projects build on this base, aiming to add further value to the project or to meet changing regulatory or HSE standards. Incremental projects that change the production profile fall into two categories: acceleration projects that add value by bringing forward production, and additional recovery projects that increase the remaining reserves. Late-life incremental projects can assist in deferring decommissioning.

Chapter 9 recommends evaluating any incremental project proposal by subtracting the net cash flow of the reference case model (no further activity, NFA) from the composite case that includes all elements of the incremental project that would not have otherwise occurred. The difference will be the true value of the incremental project.

Incremental projects that require limited capex, such as well workovers, often have a short payback time and a high IRR, and present attractive opportunities. While the NPV of each modest incremental project may be limited, multiple repeats of similar activities can create an attractive business. Significant capex may be required for larger incremental projects such as adding compression to a gas field, or applying enhanced oil recovery techniques to an oil field.

As for any investment project, it is competing for capital expenditure, and Chapter 9 introduces criteria for ranking incremental opportunities. The criteria include considerations in addition to the economic indicators introduced in Chapter 7, such as safety and regulatory requirements.

1.8 Decommissioning

When the project net cash flow turns negative, as in Year 9 of the example in Fig. 1.2, the project has reached its EL and decommissioning of the infrastructure is required. This may occur sooner than the EL if the facilities fail to meet technical integrity requirements, in which case the technical limit determines the point at which the operator declares the Cessation of Production (CoP). Unless the operator can find a solution to extend the CoP date, then the costly undertaking of decommissioning of wells and facilities will commence.

Chapter 9 describes the typical activities that are required to decommission wells, production facilities and pipelines, with reference to guidelines and legislation developed to cover the United Kingdom Continental Shelf, which is one of the regions with more experience of decommissioning facilities, along with the Gulf of Mexico. Decommissioning costs are reducing as the technology advances and experience builds up within the industry, with specialised service companies performing many of the decommissioning activities.

The owners of the asset rely on net cash flow from other projects or savings that have been set aside in a decommissioning fund to meet the liability. Chapter 9 demonstrates how decommissioning expenditure may be tax deductible on a ring-fenced loss carry back basis. It explains that commercial arrangements such as decommissioning security agreements act to safeguard governments and prior owners against current owners who fail to meet the cost liabilities.

1.9 Portfolio investment

The final chapter of the book considers the benefits of combining individual E&P assets into a portfolio. While we are familiar with the phrase 'do not put all your eggs into one basket', Chapter 10 discusses why the phrase is true, and shows how reduction in financial risk can be achieved with a combination of assets.

The financial adviser's definition of risk, being the standard deviation of value divided by the mean value (σ/μ), is used throughout this chapter, and risk-reward plots are used to express the mean value (the reward) of a particular combination of assets within the portfolio versus the risk.

When individual asset values are linked by a common external factor, such as commodity price for E&P assets, a covariance exists between the asset values, and so-called simple diversification will only reduce portfolio risk down to an undiversifiable, residual risk or systematic risk. A method commonly used in financial planning to improve on simple diversification is Markowitz diversification, which relies on investing in a portfolio of assets that have a weak correlation between them. Chapter 10 shows how this can be applied in the E&P context.

For each alternative portfolio choice, the risk and reward is calculated and placed on a risk-reward plot. The locus of the efficient frontier is drawn on the plot, as a line connecting portfolios with optimum risk-reward balances. Any portfolio that is not located on the efficient frontier is sub-optimal. Selecting where to be along the efficient frontier is based on the corporate strategy and the corporate utility function, but the efficient frontier provides a technique for choosing the optimum portfolio that either delivers the target value at minimum risk, or delivers the maximum value for a target level of risk.

Feasibility of portfolios is often constrained by factors such as capital available, and investors should understand their capacity to withstand a series of losses. Chapter 10 shows examples of the benefits of holding a diversified portfolio of E&P assets, using methods that can be performed with spreadsheets and Monte Carlo simulation.

1.10 Abbreviations and units

Appendix 1 carries a list of abbreviations and units used in the book.

Reference

[1] F. Jahn, M. Cook, M. Graham, Hydrocarbon Exploration and Production, second ed., Elsevier, 2008.
[2] Johan Svedrup is in production, Equinor, August 2020. https://www.equinor.com/en/what-we-do/johan-sverdrup.html.

CHAPTER 2

Trends in global energy supply and demand

Contents

2.1 Supply and demand curves for general commodities	15
2.2 Historical trends in global energy supply and demand	20
2.2.1 Immediate impact of Covid-19 on energy demand	21
2.2.2 Historical oil supply	23
2.2.3 Historical gas supply	26
2.2.4 Gas sales agreements and influence on gas sales price	31
2.3 Forecasts of global energy supply and demand	33
2.3.1 Overall energy demand forecasts and role for renewables	34
2.3.2 Oil demand forecasts	37
2.3.3 Gas demand forecasts	39
References	42

2.1 Supply and demand curves for general commodities

Commodities such as agricultural products, metals, minerals, oil and gas are subject to large price fluctuations in response to the balance of supply and demand, but oil and gas prices are particularly volatile. This chapter will consider the historical supply and demand for oil and gas, and refer to forecasts made by selected professional groups.

Oil and gas are both primary commodities, meaning that they are unprocessed raw materials (primary) and are essentially the same product everywhere in the world (commodity). A market provides a set of arrangements by which buyers and sellers exchange goods. The exchange is transacted by intermediaries, such as the London-based International Petroleum Exchange (IPE), or the Atlanta-based Intercontinental Exchange (ICE). The market allows buyers and sellers to determine an equilibrium price at which the quantity of goods the buyer wishes to purchase matches the quantity that the seller is willing to produce.

In 1890, Alfred Marshal wrote Principles of Economics, in which he developed a supply and demand relationship that explained the response of suppliers and buyers to changes in the price of goods. A classic supply and demand graph, a feature of nearly every subsequent economics textbook, is shown in Fig. 2.1. The x-axis represents the quantity of a good, and the y-axis the price of that good. A good is a general term for a physical commodity such as steel, gold, oranges, as opposed to services such as haircuts or financial advice. Goods can be transported, stored and exchanged, whereas services are only used at the moment of production. Oil and gas are both goods.

The solid lines in Fig. 2.1 are the supply and demand curves, which happen to be straight lines in this example, and represent the behaviour of sellers (suppliers) and consumers (buyers) to the price of the good. The supply curve (S_1) is the quantity of a good that suppliers are willing to produce at any given price. The demand curve (D_1) is the quantity of the good that the buyer is willing to purchase at any given price. It is important to realise that these are behaviours that show the relationship between price and quantity at a given point in time, when all other factors are held constant, or, as economists would say, 'all other things being equal'. For the demand curve, the main other factors that are assumed to be constant are consumer taste, the price of substitute goods and the income of buyers. For the supply curve, the constants are assumed to be technology, production costs and regulatory requirements.

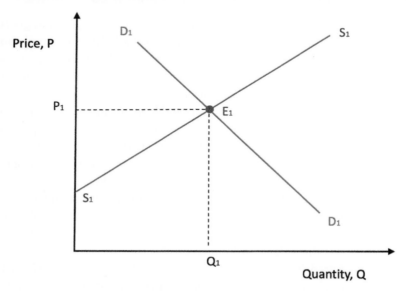

Figure 2.1 Supply and demand curves showing an equilibrium.

When a market allows buyers and sellers to trade a good freely, then an equilibrium will become established (E_1) whereby the supply meets demand at a certain quantity of the good (Q_1) which is traded at an equilibrium price (P_1). The equilibrium becomes established because, with all other things being equal, if the suppliers produce more than the buyers demand (excess supply), there will be excess of goods in the market, the supplier will build up stocks, and subsequently reduce price to clear stock. On the other hand, if there is more demand than suppliers can meet (excess demand), the suppliers will bid up the price of the good, gradually eliminating excess demand until the equilibrium is established. This description involves movements along the supply and demand curves in response to price.

When the underlying factors that were assumed equal in explaining Fig. 2.1 do change, then rather than moving along the supply and demand curves, the curves will shift, as illustrated in Fig. 2.2.

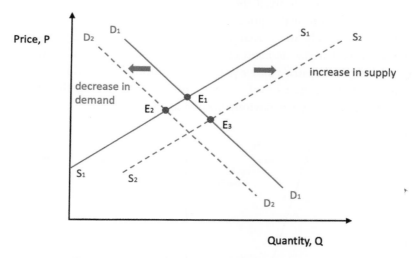

Figure 2.2 Shifts in the supply and demand curves.

The demand curve will shift to the left, to D_2, if, for example, consumer taste changes such that the good has become less appealing, or if substitute goods become cheaper, or if the buyers' income reduces. Examples in the E&P sector that would shift the demand curve to the left are a preference for cleaner sources of energy (taste), lower prices of renewables (cheaper substitutes) and lower income of buyers who economise on their use of energy (income). The shift in the demand curve in Fig. 2.2 moves the equilibrium to E_2 — a lower quantity traded at a lower price. The converse of any of these changes would shift the demand curve to the right.

The supply curve will shift to the right if technology makes the good cheaper to produce, or operating costs can be reduced, or if regulations are relaxed. Examples in the E&P sector would be improved fracking methods for tight reservoirs which delivers more hydrocarbon production for the same cost (technology), or reduction of transportation costs, perhaps due to cheaper fuel (operating costs), or lifting of trade sanctions that have been limiting oil export (regulation). The shift in the supply curve in Fig. 2.2 moves the equilibrium from E_1 to E_3 — a higher quantity, but traded at a lower price. The converse of any of these changes would shift the supply curve to the left, for example, if new government regulations were imposed, such as a carbon emissions tax, or tightening any legislation that increases operating cost.

In general, the shifts in either the supply or demand curves are caused by external factors, rather than adjustments made in response to changing price, which would be movements along the curve for either supplier (who is willing to produce more if the price is higher) or the consumer (who is willing to buy more if the price is lower).

While the supply and demand curves for oil follow this general logic, a key feature is that the demand curve for oil is very steep, as illustrated in Fig. 2.3. This is because, under general conditions, the world demands an almost constant amount of oil, with a weak dependency on the price. Oil demand is displaying a behaviour known as price inelasticity.

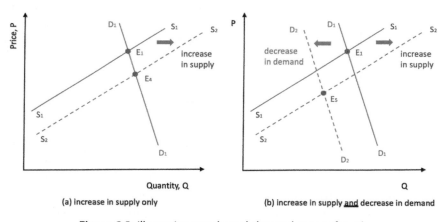

Figure 2.3 Illustrative supply and demand curves for oil.

In 2018, world oil production averaged 95.2 million barrels per day (MMb/d), and 95.3 MMb/d in 2019 according to the BP Statistical Review of World Energy 2020 (Ref. [1]). These figures include condensate and natural gas liquid (NGL) produced in association with gas. World consumption figures are very similar, with the small differences being due to changing inventory levels in storage tanks. This demonstrates that in normal circumstances world oil demand is relatively constant in the short term.

Fig. 2.3(a) illustrates that an increase in supply shifts the supply curve to the right and the equilibrium point drops from E_1 to E_4, precipitating a significant drop in the oil price for a relatively modest change in supply.

In the period 2010 to 2019, US crude oil production increased from around 5 MMb/d to 12 MMb/d as a result of the acceleration of shale oil production. The technology advances in drilling, fracking and completing the onshore wells reduced per barrel production costs and provided the incentive for the production companies to increase supply. This was one of the main contributors to a significant reduction on oil price over that period, particularly from 2014, as evident in Fig. 6.3. Other contributing factors influencing supply during that period included the easing of international sanctions on Iran and a recovery in Libyan oil production.

The assumption that global demand for oil is relatively constant in the short term was tested in the first quarter of 2020. At the time of writing, the world economy is adjusting to the effects of the Covid-19 pandemic, following the first diagnosis of the disease in China in December 2019. By the end of December 2020, more than 1.7 million people had died and 77 million cases of infection had been recorded worldwide. The lockdowns imposed to restrict the growth of the disease created unprecedented changes in energy demand. By March 2020, the demand for oil had dropped by 28 MMb/d from end-2019 levels. This was an unprecedented reduction in world demand for oil, coming at a time of record high supply.

Fig. 2.3(b) illustrates the amplified impact of a simultaneous reduction in demand and increase in supply when the demand curve is steep. The combined effect of an increase in supply and a decrease in demand moves the equilibrium point from E_1 to E_5. Oil price dropped to a historic low in March 2020, with negative oil prices for a short period as suppliers struggled to find storage and offload cargoes of crude oil. In the short term, a situation known as contango emerged.

In contango, the futures (or forward) price of a commodity is higher than the spot (current) price. Some countries and companies with storage capacity were able to purchase crude oil at a very low spot price, with the intention of selling it later, either directly or as part of a futures contract. Storage facilities included conventional stock tanks, crude oil cargo vessels and even underground storage in disused salt caverns. For those companies with storage options, the profit made by their trading arm from taking advantage of the contango partially offset the impact of generally lower oil prices in 2020. Some further observations on the immediate impact of Covid-19 are made in Section 2.2.1.

While the logic of supply and demand curves has a solid foundation, built on the work of Marshall, they are not the only factors influencing oil price in the short term. The attitude of traders in the markets plays a role in setting spot prices and short-term future contract prices, and their sentiment and actions can be influenced by their perception of future conditions, rather than the direct logic of supply and demand forecasts.

2.2 Historical trends in global energy supply and demand

BP produces a report, the Statistical Review of World Energy, which has been released every year since 1952 and is freely available online (Ref. [1]). The review presents analyses of world energy markets from the previous year, and tracks historical trends for many forms of energy including fossil fuels, nuclear, hydroelectricity and more recently renewable energy. It is based on a compilation of data from multiple sources, and is widely respected within the energy industry as being reliable.

The left-hand graph in Fig. 2.4 shows a steady increase in global energy demand from 1994 to 2019, which was driven by global population growth, and the industrialisation of developing economies, particularly in China, India and on the African continent. The dip in 2009 was a result of the global financial crisis, and another aberration will appear from 2020 as a result of the Covid-19 pandemic.

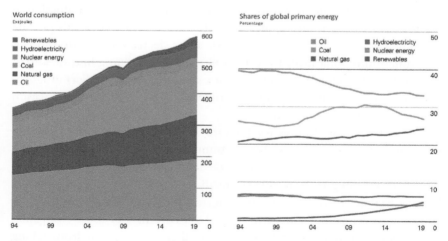

Figure 2.4 Primary energy consumption and sources. *(Source: BP Statistical Review of World Energy 2019, reproduced with permission of BP.)*

The traditional fossil fuels (oil, gas and coal) represented 84% of the sources of primary energy in 2019, but some significant trends are apparent in the right-hand graph. While the demand for oil increased steadily for 25 years in absolute terms, its share of the supply to meet global energy demand reduced, partly because of the increased focus on gas as a cleaner fuel, and partly due to the increase in energy supply from renewables. While coal usage increased in the 1990s and early 2000s, partly due to intense building of coal-fired power stations in China, this has now reduced through moves to cleaner forms of power. Nevertheless, in 2019 coal was the single largest source of electrical power, used to generate 36% of global electricity. The UK Government confirmed in 2020 that its remaining coal-fired power stations would be shut by 2025, as part of a switch to gas and renewable sources, particularly wind power.

Growth in global energy demand in 2019 was slower than in 2018 due to weaker global economic growth, and the impact of Covid-19 will slow this further in 2020.[1]

2.2.1 Immediate impact of Covid-19 on energy demand

In the six months after the first diagnosis of Covid-19 (December 2019), the ensuing pandemic had a major impact on global energy demand as countries locked down people and businesses restricted their activities in an effort to contain the spread of the virus. The health, social and economic impact was on a scale not seen since World War II.

The demand for energy fell sharply, as transport, business and economies slowed. By March 2020, the number of global commercial flights had reduced by 80% compared to the beginning of 2020, and road traffic levels reduced by more than 50% in many countries compared to the levels 12 months previously. Road traffic data was collected from sources including TomTom Traffic Index and Google Maps, and collated by Rystad Energy (Ref. [2]). At the low point of oil demand in March 2020, the level had dropped by some 28 MMb/d, compared to end 2019 demand, representing a 29% decrease.

In the immediate term, with no cutbacks in production, an over-supply of crude oil sent oil prices tumbling to historic lows as stock tanks reached capacity, and oil was unable to be offloaded from tankers. In March 2020 West Texas Intermediate crude oil (WTI) dipped to -$40/bbl (minus!), as

---

[1] The unit of measure of energy in Fig. 2.4 is Exajoule. For reference, 1 Exajoule = 10^{18} J, 1 bbl oil = 6.12×10^9 J and 1 scm of natural gas = 36×10^6 J.

buyers were effectively being paid to accept crude shipments. This anomaly was addressed by an agreement between the Organization of the Petroleum Exporting Countries plus Russia (OPEC+) to cut back production by some 9 MMb/d by June. This rebalanced the market to some extent, with crude oil prices stabilising around $40/bbl in July 2020. Saudi Arabia and Russia took approximately half of the total cut, and compliance to targets set within the OPEC + group was generally observed. Oil production in the United States fell by some 1.6 MMb/d as wells were either shut in or awaited fracking, but started to recover somewhat as the oil price stabilised. In total, non-OPEC + countries (including the United States) shut in approximately 3 MMb/d in June.

In the first quarter of 2020, oil majors announced large cutbacks in capital expenditure for the year, and reported significant losses for the second quarter of the year. The International Energy Agency (IEA) estimated that capital investment in the oil and gas sector would be some 32% lower in 2020 than in 2019, and that during 2020 global consumer expenditure on electricity would be greater than the expenditure on oil for the first time.

The longer-term post-Covid-19 recovery of demand for energy and its impact on oil production and price are topics of great uncertainty at the time of writing. Many analysts, including Rystad and the IEA, carry scenarios for recovery of energy demand which span from V-shaped (rapid recovery), through U-shaped (slower recovery) to L-shaped (permanent downturn). The V-shaped recovery forecasts assume a resumption of former behaviours, while L-shaped forecasts assume a change of behaviour of business and society, such as more working from home, less international travel and improved business efficiency.

As governments struggle to recover from the impact of Covid-19, options exist to restructure the energy supply and infrastructure to boost their economies, create jobs and address greenhouse gas emissions. The IEA report Sustainable Recovery (Ref. [3]) analyses the options available at a time of unique opportunity, and proposes a sustainable recovery plan to policy makers, industry and investors. This includes a plan for the energy sector and recommends an acceleration of electrification and a shift to low-carbon energy sources. While the plan is tailored to global regions, common themes include expanding electricity distribution grids, increasing use of solar and wind energy for electricity generation, modernisation of hydropower and nuclear plants, reduction in industrial methane emissions and improved efficiencies in transport systems and building practices. Investment in these areas is predicted to generate an increased gross domestic product (GDP) globally, but most markedly in less advanced countries.

It is not yet clear what impact Covid-19 will have on the development of renewables. Several factors are likely to influence this. The drop in oil price and ongoing cheap gas make renewables less attractive. On the other hand, the immediate improvement in cleaner air from the reduction of transportation, as observed in the first six months of 2020, is a demonstration of a benefit of reducing fossil fuel consumption, particularly in densely populated urban areas. In New Delhi, NO_2 (nitrous oxide) levels dropped by 66% during the first 2 weeks of lockdown as traffic congestion eased. The consequent benefits to health may accelerate policy decisions in favour of accelerating the development and implementation of renewables.

Stepping back to energy supply in 2019, oil and natural gas provided some 57% of world primary energy, and the following section will summarise where the supply lies, and how long it would last at the current rate of production.

2.2.2 Historical oil supply

The current supply of hydrocarbons comes from proved reserves, which are distributed globally as shown in Fig. 2.5. The term proved reserves has a specific technical meaning, in that it represents volumes that geological and engineering information indicate with reasonable certainty can be recovered in the future from known reservoirs under existing economic and operating conditions. If estimated using probabilistic methods, proved reserves have approximately 90% chance of being recovered. There is a

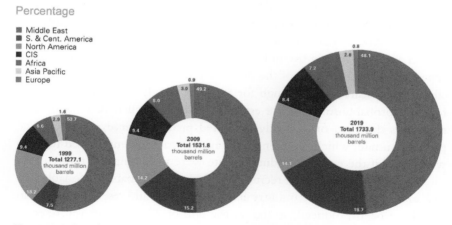

Figure 2.5 Distribution of proved oil reserves 1999, 2009 and 2019. *(Source: BP Statistical Review of World Energy 2019, reproduced with permission of BP.)*

significant upside to proved reserves from the probable and possible reserves classes. Classification of reserves is described in Section 3.4, and methods of estimating the proved reserves are discussed in Section 4.2.2.

The pie charts in Fig. 2.5 represent the world's total proved oil reserves over two decades, with the percentage of the total reserves in each region indicated in the sectors. The size of the pie is not to scale with the total proved reserves, shown the centre of each pie. It is clear that over the two decades represented, the E&P industry has been very effective at adding proved reserves, despite the interim average production of around 300 thousand million barrels per decade.

The distribution of proved reserves is, and has been for decades, in favour of the Middle East, where oil is largely onshore in easily accessed, high quality and productive reservoirs. By 2019, Central and South America took second place, following significant offshore discoveries and developments in Brazil in the early 2000s. North American reserves have been boosted through the development of onshore shale oil, particularly in the Permian basin straddling West Texas and New Mexico, and in the Bakken shale in North Dakota. Commercial production of oil from tight shale oil reservoirs has been enabled by the application of fracking technology, and increased efficiencies in drilling and completing wells. By 2019, the United States had become the world's highest rate producer, averaging 17.0 MMb/d of total liquids, and leading Saudi Arabia (11.8 MMb/d) and Russia (11.5 MMb/d). Note that total liquid figures include the production of condensate and NGLs recovered from wet gas production during processing. Oil production alone peaked at 12.3 MMb/d in the United States in 2019. The proved reserves in Fig. 2.5 also refer to total liquids. All production numbers are taken from Ref. [1]. CIS is the Commonwealth of Independent States (Azerbaijan, Armenia, Belarus, Georgia, Kazakhstan, Kyrgyzstan, Moldova, Russia, Tajikistan, Turkmenistan, Uzbekistan and Ukraine).

On applying to join the industry in 1981, I asked how long oil would last — the natural concern of the new recruit. The answer I was given was 'about forty years', long enough for a career, which was comforting. The recruiter probably founded the estimate on the so-called R/P ratio. This represents the ratio of currently estimated proved reserves to the annual production rate, which represents how many years current proved reserves would last, if the current production rate were maintained in future.

Fig. 2.6 illustrates that the recruiter's forecast was probably reasonable, but that on a global basis the industry has been able to add reserves faster than they have been produced, so that at the end of 2019 the proved reserves would last for 50 years if production were to be maintained at the end-2019 rate.

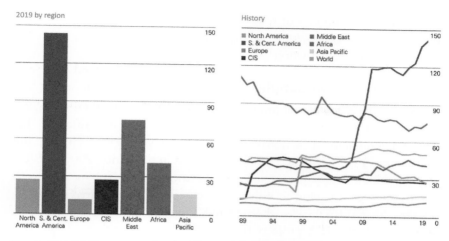

Figure 2.6 Oil Reserves-to-Production ratio (R/P) forecasts by region. *(Source: BP Statistical Review of World Energy 2019, reproduced with permission of BP.)*

Some regional R/P ratios stand out. South and Central America's R/P ratio is anomalously high because the major projects that carry proved reserves have either started production recently, or are awaiting first production. Discounting that anomaly, the Middle East has by far the highest R/P ratio, consistent with the proved reserves shown in Fig. 2.5. Africa has shown significant growth over two decades, with both onshore and offshore developments. It is apparent that the industry is not lacking in the opportunities or skills to make oil discoveries and bring them into production.

A similar R/P calculation can be performed for an individual company, to consider its future dependence on current proved reserves. However, it is more common to consider the company's annual performance by calculating the reserves replacement ratio (RR). This is the proved reserves added during the year divided by the total annual production, and would be 1.0 if the company added the same amount of proved reserves as it produced, representing a sustainable business. The RR ratio is often used as a key performance indicator (KPI) in setting targets for the exploration and development team, and by analysts in gauging the future security of the business. The RR target may be relaxed by using the 2P (proved plus probable) reserves class. A version of this ratio may also be applied purely to exploration by comparing volumes discovered by the exploration team to annual production. As a detail, exploration discoveries would add volumes of contingent resources, not reserves, since reserves must be demonstrated to be economic to develop. For the definitions of 2P reserves and contingent resources, see Section 3.4.

2.2.3 Historical gas supply

As Fig. 2.4 shows, while oil production has declined over two decades, gas production has increased. During this time, increased electrification has been implemented at national and regional levels. For example, the International Energy Agency (IEA) reports that in the period 2000 to 2019, almost 1 billion people gained access to electricity in developing Asia, with 94% of the region having access to electricity in 2018 compared with 67% in 2000 (Ref. [4]). Under their Stated Policies Scenario, explained in Section 2.3.1, the IEA projects that by 2030 the electrification rate will reach 99% of the population of developing Asia. Strong current advances in electrification are also noted in East Africa. The electrification of transport systems, both public and private, is also a significant development.

This rising demand for electricity and the high efficiency with which gas generates power have driven the demand for gas and shifted the focus of exploration and development in many E&P companies, particularly the oil majors. At the end of 2019, the relative share of proved reserves for Shell, as reported in the 2019 Annual Report (Ref. [5]), was 44% oil and NGLs and 56% natural gas. This includes proved developed (PD) and proved undeveloped (PUD) reserves from all Shell subsidiaries and Shell's share of Joint Ventures and associates. The percentages are based on the total barrels of oil equivalent (boe) of oil and gas using a conversion of 5800 scf gas equivalent to one barrel of oil. The split for BP, on the same basis, was 59% oil and 41% gas, at the end of 2019 (Ref. [6]) and for ExxonMobil 64% oil and 36% gas (Ref. [7]). These companies are known as international oil majors, but a very significant fraction of their portfolio is in gas.

Gas is a significantly cleaner fossil fuel than coal for generating electricity, emitting around 50% of the CO_2 emissions per unit of electricity, making gas a key fuel in the transition towards a net-zero carbon economy. Net-zero carbon has become the target of many organisations such as the European Union (EU), individual countries and some companies, such as BP, Shell and Barclays. These stated ambitions are in response to the 2015 Paris Agreement, which set a global goal to reach net-zero emissions in the second half of the century.

Fig. 2.7 shows the global distribution of proved gas reserves across two decades to 2019, in a similar format to Fig. 2.5. Over those 20 years, proved gas reserves have increased by 50%, despite interim production. Proved oil reserves increased by 36% over the same 20 years (Fig. 2.5), reinforcing the shift in focus of the E&P industry toward gas. As for oil, a very significant fraction of the gas reserves lie in the Middle East, a region that has traditionally concentrated on oil production. A third of the world's gas reserves lie in the CIS, mostly in Siberia, which exports much of its production to Western Europe, and most recently to China. The Middle East and Russia dominate the proved gas reserves at end 2019, but large discoveries in East Africa (Mozambique, Tanzania) and South America (Brazil, Venezuela, Trinidad and Tobago) will become apparent once projects are sanctioned and volumes become classed as reserves.

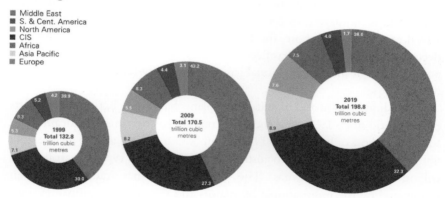

Figure 2.7 Distribution of proved gas reserves 1999, 2009 and 2019. *(Source: BP Statistical Review of World Energy 2019, reproduced with permission of BP.)*

According to BP's Statistical Review of World Energy, the world's R/P ratio for gas at the end of 2019 is 50 years, and for the Middle East region 109 years and CIS 60 years. The Middle East R/P has been steadily reducing since the turn of the century as gas fields have been commercialised. The volumes in Fig. 2.7 are presented in trillion cubic meters, and if converted to boe, then proved gas reserves at end 2019 are approximately 1200 thousand million boe, which can be compared with the 1734 thousand million barrels of proved oil reserves. Note again that the oil reserves include the NGLs and condensate recovered from wet gas production during gas field development.

Transportation of gas to markets is expensive compared to oil distribution because of the relatively low energy density of gas. It is more economical to sell gas close to its point of production. To reduce gas transportation costs on land, an economy of scale can be achieved by transporting at high rate in large pipelines. For example, the Power of Siberia gas pipeline, also known as the China—Russia East-Route Natural Gas pipeline, which started gas export at the end of 2019, is nearly 4000 km long, has a diameter of 1.42 m (56 inches) and has a throughput capacity of 61 billion cubic metres per annum, or approximately 3.6 billion scf/d. For context, the UK natural gas consumption in 2019 is approximately 7.5 billion scf/d.

To transport gas large distances across seas, and to maintain acceptable economics, the dry gas (methane) production from the gas field is compressed and chilled to form liquefied natural gas (LNG). This reduces the gas volume by a factor of 600, which greatly improves the energy density. The LNG is transferred to refrigerated storage tanks on LNG tankers. On arrival at a deepwater port, the LNG is offloaded to a facility where it is decompressed to form regasified methane, which can then be fed into power plants or into conventional gas distribution systems for industrial and domestic use.

LNG liquefaction and regasification plants have been built in many parts of the world since 1990, and the LNG market has become significant. Due to the high cost of building an LNG plant ($2—5 billion) a large recoverable volume of gas is required to justify an LNG project, typically around 10 trillion scf (10 Tscf) of gas. The LNG market has traditionally been dominated by the international oil majors and national oil companies with the financial strength to undertake these capital-intensive projects. However, developments of smaller scale LNG projects are now emerging, opening up the market to mid-sized companies and private enterprises.

Technology has been developed during the early 2000s to liquefy produced natural gas on marine vessels, a system known as floating LNG (FLNG). The FLNG vessel is located at the development site, and offloads its product to a tanker that transports the liquid to an LNG terminal facility. At the end of a gas field lifetime, the FLNG vessel can be redeployed to other gas field developments.

FLNG technology has driven down the reserves threshold for economic development of offshore gas fields to around 1 Tscf. This can provide a lifeline to relatively small offshore gas discoveries that would otherwise remain undeveloped, a frustrating status known as stranded gas. At the time of writing, the largest FLNG development is that undertaken by Shell who commissioned the construction of the 488-m-long Prelude FLNG vessel, built in South Korea at a cost of some $14 billion, for development of a remote gas field in the Browse Basin, offshore Western Australia (Ref. [8]). The vessel moved onto location at the end of 2017, and started production a year later. In full operation it will produce 3.6 million tonnes of LNG per annum (4.9 billion cubic metres of natural gas), plus condensate and LPG (liquefied petroleum gas, a mixture of propane and butane), being a total of 100 thousand bbl/d of liquids. Other major FLNG projects include the Petronas PFLNG 1 and 2, offshore Sabah, Malaysia and Eni's Coral FLNG vessel offshore Mozambique.

Transportation of LNG across large distances has opened up an active global market. The economics of the technology has allowed the United States to start to export gas from the tight onshore gas fields to a world market. Fig. 2.8 shows the volumes of LNG imported into selected regions. Japan, having no indigenous oil and gas resources, has relied on LNG import for several decades, but note the increase in LNG import into China between 2005 and 2019 to support its industrial growth. Since 2015, the United States has been exporting LNG using plants originally designed for regasifying LNG imports to secure energy supply, but then reconfigured into liquefaction plants for LNG export.

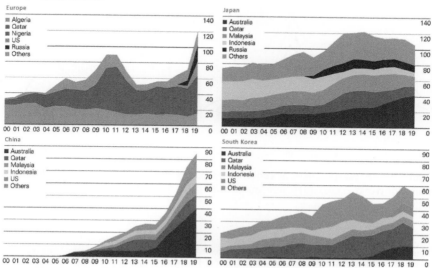

Figure 2.8 LNG imports by source. *(Source: BP Statistical Review of World Energy 2019, reproduced with permission of BP.)*

The worldwide movement of gas in the form of LNG, with relatively low transportation costs, has had the effect of moving gas towards being a commodity that can be globally traded, rather than sold to local market. This has tended to normalise the price of gas as shown in Fig. 2.9, in which the scale is in dollars per million Btu (British thermal unit).[2]

US gas prices have been supressed for the decade to 2019 due to abundant domestic production, particularly from the development of tight gas. Production in excess of domestic demand has provided the opportunity to export LNG, as shown in Fig. 2.8. The impact of the flexibility in destination and a developing spot market for the sale of LNG has led to more uniformity in gas sales prices. The effect in Japan halved gas prices in the decade to 2019. There are similarities between the impact of growth in US shale oil production on oil prices, and the effect of increased US tight gas production on gas prices, coupled with the increased transportability of LNG. For both oil and gas, the result has been to lower the price of the commodity.

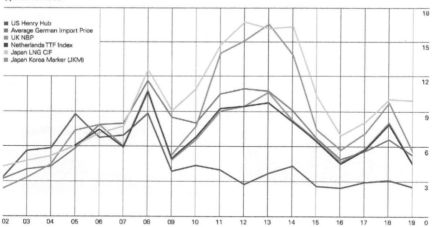

Figure 2.9 Global gas prices at various sales points 2002 - 2019. *(Source: BP Statistical Review of World Energy 2019, reproduced with permission of BP.)*

---

[2] One million Btu is the energy contained in approximately 1000 standard cubic feet of gas, depending on its composition. The gas sales points (e.g. US Henry Hub) referred to in the legend in Fig. 2.9 are explained in Section 6.2.

2.2.4 Gas sales agreements and influence on gas sales price

Uncertainty in the gas sales price for production from a specific project can be reduced by agreeing to long-term gas sales contracts with a buyer, typically for a term of 15—25 years, and this will avoid the gas sales price fluctuations of the spot market shown in Fig. 2.9. The buyer may be an intermediary between the producer and the final consumer, buying the gas at a delivery point and taking responsibility for the marketing and distribution.

A gas sales agreement (GSA) is signed between the producer and the buyer, often at an early stage of the project life, even pre-sanction, to provide comfort to the investor that the project revenue stream will deliver acceptable economic targets. GSAs fall into two general categories. Depletion Contracts cover the production from a specific field whose total production is usually dedicated to a single buyer, with agreed quantities of production linked to the build-up, plateau and decline periods of the field lifecycle. The second category includes Supply Contracts in which the seller agrees to delivery quantities without specifying the source of the gas, which provides the seller with the flexibility to produce from any of the fields within their portfolio, as long as product specifications are met.

In all GSAs, the quality of the natural gas delivered must meet specifications on calorific value and maximum levels of impurities (water, CO_2, hydrogen sulphide (H2S), hydrocarbon liquids). It is the responsibility of the producer to meet these through adequate processing and treatment of production from the gas wells.

Some of the common terms within GSAs are shown in Fig. 2.10. This uses an example of a Peak Supply Contract in which there is a seasonal variation in the demand for gas, shown as the red line, as this allows the description of key terms found in GSAs.

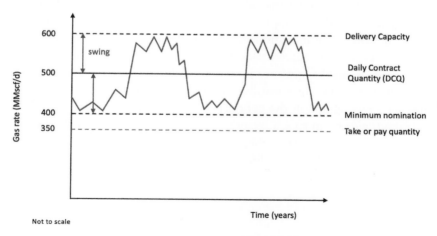

Figure 2.10 Typical components of a long-term gas sales agreement.

The Daily Contract Quantity (DCQ) is the agreed daily quantity of gas, which will generally be fixed for a Supply Contract, but will change in a Depletion Contract for the different stages of the field lifecycle. The DCQ is a volume term (MMscf or MMm^3), usually expressed as a daily rate (MMscf/d or MMm^3/d). The Annual Contract Quantity (ACQ) is the amount of gas the buyer expects to take in a Contract Year, which is the DCQ multiplied by the number of days in the year.

The DCQ forms the foundation of the quantity to be produced, chosen to be 500 MMscf/d in Fig. 2.10. The example shows that the quantity the buyer asks for (nominates) varies with season, and even varies on a daily basis as ambient temperature changes and end users demand more gas for heating as temperatures drop or more air conditioning as temperatures rise. The Delivery Capacity is the agreed maximum daily quantity that the buyer can nominate. It is also the maximum daily quantity that the seller is obliged to deliver, and this is translated into a daily rate of 600 MMscf/d in the example.

The difference between the Delivery Capacity and the DCQ is known as the 'swing', which is 100 MMscf/d in the example. The GSA will include an agreed swing, which is usually constant throughout the contract, but in some cases may vary with the season. The higher the swing, the higher the agreed gas sales price will be for all gas sold. To achieve the peaks of production, nominated at short notice, perhaps on the preceding day, the producer must have the potential deliverability of the gas wells, and the capacity and availability in the processing plant and pipeline to deliver the quantity demanded. This requires incremental investment compared to planning only for the DCQ, and very careful management of reservoir, wells and facilities during operations. The expense of incremental effort is rewarded by a higher gas sales price.

Importantly, the seller may also nominate the DCQ minus the swing, 400 MMscf/d in the example, known as the minimum nomination. As long as the buyer's nomination lies within the minimum nomination and delivery capacity, the agreed sales price for each unit of gas applies. If the buyer does demand a quantity in excess of the delivery capacity, and the seller can meet this, then usually a premium for the excess gas sales will apply, at a rate agreed in the contract.

If the buyer nominates a quantity below the delivery capacity, but the seller is unable to meet that obligation, then a penalty is applied based on the 'shortfall', which is the difference between the nomination and the (from WEO 2019), which is based on STEPS. The green dots in Fig. 2.12 indicate the share of the 2019–40 additions to global power generation that renewables would have to make to achieve the targets assumed in the SDS.

In Fig. 2.12, the 'other renewables' category includes geothermal, concentrating solar power, bioenergy and marine energy. Concentrating solar power technology converts sunlight into heat by use of lenses and mirrors. The heat is used to raise steam which drives turbines for electricity generation. C and S America is Central and South America. In EIA terminology, low-carbon sources include all renewables above, plus hydrogen and nuclear energy.

Under the STEPS scenario, solar PV sources are forecast just to exceed gas as a source of global power generation by 2040. Under SDS assumptions, solar PV would generate twice as much power as gas by 2040, while wind would also exceed gas. The SDS represents major further investment in solar PV and wind. WEO 2019 notes the huge potential for solar in Africa, where it has the opportunity to provide the continent with more than its forecast electricity needs. Again, this requires commitment to major investment, which has so far been slow, and lack of access to electricity remains an acute barrier to progress for people in sub-Saharan Africa.

2.3.2 Oil demand forecasts

While the further development of renewables, particularly solar PV and wind, is a key component in meeting increased global demand for energy and imperatives aimed at global environmental protection, oil and gas continue to contribute significantly in all WEO 2019 scenarios, as Fig. 2.11 shows. However, under the STEPS scenario, there are significant geographical changes in supply and demand for oil between 2018 and 2040, as shown in Fig. 2.13.

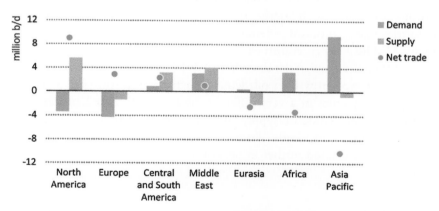

Figure 2.13 Geographical change in oil supply and demand under WEO 2019 STEPS scenario, 2018-2040. *(Source: IEA (2019), World Energy Outlook. All rights reserved.)*

In Fig. 2.13, supply and demand are the change in supply from within the region and change in demand within the region, in the period between 2018 and 2040. Positive net trade values are increases in net exports, negative values are increases in net imports. Oil demand in the Asia Pacific region increases significantly as supply drops marginally, resulting in a marked shift in global oil trade towards Asia. Europe's oil demand drops faster than the drop in supply, allowing it to have a positive net trade.

WEO 2019 STEPS forecasts overall growth in global oil demand, reaching 106 MMb/d by 2040, with US tight oil production growing from 6 MMb/d in 2018 to 11 MMb/d in 2035. The increase in US output to 2030 accounts for 85% of the global increase in oil production, with the United States becoming the world's second-largest oil exporter by 2030. Meanwhile US oil consumption drops, so that net trade in oil in the United States shows a significant upwards movement. This shifts the current influence away from the OPEC+ (OPEC plus Russia) producers towards the United States, as global oil trade becomes transformed.

Despite the forecast slowdown in China's oil demand from the early 2030s, China becomes the world's largest consumer just before 2040, and the Asia Pacific region demand increases significantly as India's oil demand doubles to 9 MMb/d between 2019 and 2040, the largest absolute growth of any country. In Fig. 2.13, Eurasia is Russia, while Asia Pacific region includes China, India, Japan and Southeast Asia.

Note that WEO 2019 was published prior to the Covid-19 pandemic, and therefore forecasts do not incorporate short-term effects, or any long-term implications for energy demand. The immediate impact of Covid-19 on oil price creates a challenge for regions with high production costs, such as the tight oil in the United States, and for countries that rely on oil revenues to balance their national budgets, like many Middle Eastern countries.

STEPS is only one of the three WEO 2019 scenarios. Under the Current Policies scenario, growth in oil demand increases year-on-year, while the SDS paints a picture of a very different future for oil. Fig. 2.14 shows the forecast of global crude oil supply and the impact on the price under these three scenarios. This clearly illustrates the large uncertainty in future oil price. This is discussed further in Section 6.2.

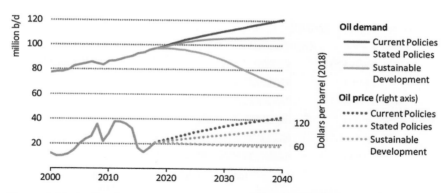

Figure 2.14 Geographical change in oil supply and demand under WEO 2019 STEPS scenario, 2018-2040. *(Source: IEA (2019), World Energy Outlook. All rights reserved.)*

2.3.3 Gas demand forecasts

WEO 2019 STEPS scenario forecasts that between 2020 and 2040 the global demand for natural gas will grow more than four times faster than the demand for oil, with developing Asian countries accounting for half of the global growth. The supply, however, comes from other regions, implying that gas will be transported over increasingly large distances to reach consumers in Asian markets. As discussed in Section 2.2.3, LNG is a key to the transportation and consequently the growth in global gas trade. Net exports from the United States will increase, alongside a net increase in imports into China.

Under this scenario, Fig. 2.15 shows the forecast of change in global gas supply in the period 2018 to 2040, measured in billions of cubic metres (bcm) per annum. By 2040, China imports almost twice as much LNG as the next largest importing country, India. Middle East growth in LNG export is led by Iraq, expanding from a low base in 2018. Other producers in the Middle East take the opportunity to diversify business away from a traditional reliance on oil production. The gathering of associated gas from oil field production is also significant to gas sales, reducing the volumes of associated gas that would once have been flared. Both domestic supply and import of gas are forecast to drop in the EU, due to increased efficiencies and strong growth in power generation from renewables.

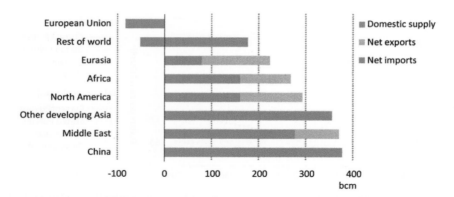

Figure 2.15 Change in annual gas supply balance by region under WEO 2019 STEPS scenario, 2018–2040. *(Source: IEA (2019), World Energy Outlook. All rights reserved.)*

A significant uncertainty in the forecasts for the development of LNG markets lies in the high cost of plant and transportation compared to alternative fuels, such as coal and the leading renewables, particularly as the production cost of energy from renewables drops as technology advances. This uncertainty creates some degree of reticence for investors in LNG, who may prefer long-term sales contracts to avoid price risk. In its favour, gas has a considerable advantage as a clean fuel compared to coal, as it produces half the CO_2 emissions than coal when producing a unit of electricity. Sulphur and particulate emissions from shipping may also be reduced by using LNG as a fuel. In a scenario including actions to meet emissions targets, LNG future demand becomes less uncertain.

In a similar vein to WEO 2019 SDS scenario, Shell's Sky scenario (Ref. [12]) presents a pathway that it believes is a technologically achievable energy solution that would meet the targets set out in the 2015 Paris Agreement on climate change. The aim of the Agreement is to hold the increase in global average temperature to well below 2°C above pre-industrial levels and to pursue efforts to limit the temperature increase to 1.5°C above pre-industrial levels. In chasing this aim, the Agreement calls for net-zero CO_2 emissions in the second half of the century. Net-zero emissions is a shorthand term for achieving a balance between man-made emissions of greenhouse gases and removals through capture and storage in sinks.

Shell's Sky scenario includes actions on consumer behaviour, energy efficiency, carbon pricing, accelerated electrification, carbon capture and storage (CCS), net-zero de-forestation and renewables producing more primary energy than fossil fuels in the 2050s. In terms of renewables, the scenario foresees solar, bioenergy and wind as the dominant contributors to power generation, with oil remaining the largest fossil fuel energy source.

Each forecaster in Table 2.1 has developed its own set of scenarios and makes significantly different assumptions about technology development, environmental and energy policies. It is beyond the scope of this book to make a succinct summary and comparison of the many forecasts, but this is presented in Ref. [10].

In summary, the world energy supply is entering a transition period. While global energy demand is expected to continue to grow under current stated policies, despite the setbacks of Covid-19, the supply of future energy is moving to cleaner fuels, renewable energy sources and electrification. Professional groups take various views of the future mix of energy supply and demand, and these views change with time. The general view is that oil, gas and coal continue to make a significant contribution to supplying world energy, and wind and solar energy currently lead the growth of renewables, which make up a significant fraction of the growth in energy demand in the 2020—40 period.

Beyond 2040, the possibility of a hydrogen-based economy may become more realistic and current investments, especially by the EU and Japan, may allow hydrogen to compete with fossil fuels (Ref. [13]). Power generation using nuclear fusion is being pursued by the $20 billion plus ITER[3] project, funded by the leading OECD[4] economies. This may transform electrical power generation during 2050—2100 and provide the planet with large-scale (2 GW power stations), safe, green, economical electrical power generation (Ref. [14—16]).

The references provided in this chapter allow the reader to keep abreast of the trends in the energy mix and the range of forecasts of future global and regional energy supply.

---

[3] ITER is an abbreviation for International Thermonuclear Experimental Reactor, also Latin for 'the way in'. It is a project involving the collaboration of 35 nations, working on magnetic fusion as a source of energy.

[4] OECD is the Organisation for Economic Cooperation and Development, made up of 37 member countries, with the aim of promoting economic growth, prosperity and sustainable development.

References

[1] BP Statistical Review of World Energy, BP. https://www.bp.com/en/global/corporate/news-and-insights/press-releases/bp-statistical-review-of-world-energy-2020-published.html.
[2] Rystad Energy Cube. https://www.rystadenergy.com/products/Data-and-Analytics-Delivery-Methods/cube-browser/.
[3] Sustainable Recovery, World Energy Outlook Special Report, International Energy Agency (IEA), July 2020. https://www,iea.org.
[4] International Energy Agency (IEA), SDG7:Data and Projections, Flagship Report, November 2019. https://www.iea.org/reports/sdg7-data-and-projections/access-to-electricity.
[5] Shell Annual Report, 2019. https://reports.shell.com/annual-report/2019/strategic-report/segments/oil-and-gas-information/reserves.php.
[6] BP Annual Report, 2019. https://www.bp.com/content/dam/bp/business-sites/en/global/corporate/pdfs/investors/bp-annual-report-and-form-20f-2019.pdf.
[7] ExxonMobil Annual Report, 2019. https://corporate.exxonmobil.com/-/media/Global/Files/annual-report/2019-Financial-and-Operating-Data.pdf.
[8] Shell Prelude FLNG project, https://www.shell.com/energy-and-innovation/natural-gas/floating-lng.html.
[9] Bansal, A.K., International Petroleum Contracts, Understanding Natural Gas Sales & Purchase Contracts and Principal Contractual Terms, February 2017 http://ashokkumarbansal.com/understanding-natural-gas-sales-purchase-contracts-principal-contractual-terms/.
[10] R.G. Newell, D. Raimi, S. Villanueva, B. Prest, Resources for the Future, Global Energy Outlook 2020: Energy Transition or Energy Addition?, May 2020. Report 20-05, https://media.rff.org/documents/GEO_2020_Report.pdf.
[11] International Energy Agency (IEA), World Energy Outlook 2019, November 2019, ISBN 978-92-64-97300-8. www.iea.org/weo.
[12] Shell Global, Shell Sky Scenario, https://www.shell.com/energy-and-innovation/the-energy-future/scenarios/shell-scenario-sky.html.
[13] J. Feder, H_2 economy: hype, horizon or here, J. Pharm. Technol. (JPT) 72 (8) (August 2020).
[14] www.iter.org.
[15] https://thebulletin.org/2019/04/fusions-greatest-hits-as-detailed-in-the-bulletin/.
[16] T. Hamacher, A.M. Bradshaw, Fusion as a Future Power Source: Recent Achievements and Prospects, World Energy Council, October 2001. http://www.worldenergy.org/wec-geis/publications/default/tech-papers/18th-Congress/downloads/ds/ds6/ds6-5.pdf.

CHAPTER 3

Gaining access to E&P opportunity

Contents

3.1 Options for accessing E&P investment opportunity	43
3.2 Strategic choices for investment	44
3.3 Bidding processes	45
3.4 Acquisition and disposal	51
3.5 Merger	55
3.6 Joint venture and partnership	57
3.7 Unitisation and equity determination	58
3.8 Financing the project	61
3.8.1 Debt seniority ranking	62
3.8.2 Debt financing	62
3.8.3 Equity financing	63
3.8.4 Mezzanine financing	64
3.8.5 Reserves-based lending	65
3.8.6 Volumetric production payments	66
3.8.7 Islamic finance	66
References	67

3.1 Options for accessing E&P investment opportunity

Many avenues are available for investing actively in the E&P business. The investor, who may be any entity ranging from a multinational to a small private company, should match the type and scale of investment to its strategy. Passive investment in the E&P sector may be made simply by buying shares in a listed company, but this chapter will focus on options available for those considering direct investment as an active participant in petroleum exploration, development or production. The techniques of petroleum economics and risk analysis covered in subsequent chapters can be applied to evaluate the potential reward and the associated risks of such investments.

In the traditional field life cycle, presented in Fig. 1.1, gaining access refers to positioning the company with the opportunity to explore for new discoveries. However, opportunities exist to enter into the E&P business at any stage of the life cycle, and this chapter will consider the practicalities of gaining access to petroleum directly through exploration, or by acquisition of producing assets, mergers with other companies, or through joint venture (JV) arrangements. In every case, finance will be required, and the chapter will summarise general financing methods available to all businesses, and those specific to the upstream oil and gas business.

3.2 Strategic choices for investment

The headings in Fig. 3.1 represent key decisions required in formulating a strategy for active investment in E&P ventures, with options listed. An investor may choose to focus on a specific target, say onshore US tight oil, or to spread the investment across different asset types and geographies. Chapter 10 will explain how a mix of asset types within a portfolio can reduce overall risk. Often the investor's preferred strategy for growing the business reflects the strengths of the organisation. For example, if the skill set is biased towards exploration, with particular experience in a geographical region or type of geology, the attention may become focussed on deepwater exploration in new basins around the world.

Another organisation may set a strategy focussed on onshore oil production, with the ambition to act as operator, targeting only opportunities that add potential of 500 MMb to the business.

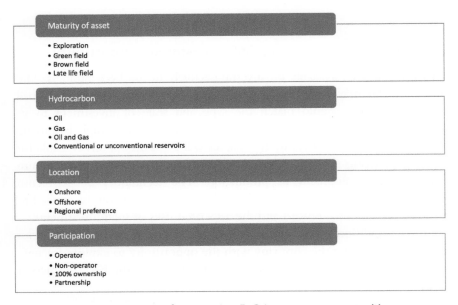

Figure 3.1 Options for accessing E&P investment opportunities.

Driven by the ongoing electrification of the energy business, some major oil companies have adjusted their strategy to balancing their traditional oil-dominated portfolio by seeking to add gas assets into the portfolio.

A clear strategy is a key guide to the appropriate method of gaining access to petroleum resources. The strategy should include a view on the level of risk that the company is willing to take, relating to its ability to withstand losses and its attitude to the risk-reward balance. This will influence not only the type of asset to seek, but also the level of participation, known as working interest, that the company wishes to undertake in an exploration or development venture.

This chapter describes the methods of accessing opportunities, starting with exploration.

3.3 Bidding processes

Countries offer opportunity for exploration by announcing licensing rounds in which blocks are opened to bidding from interested parties. A useful resource for keeping track of current licensing activity, including upcoming licensing rounds, is the *Petroleum Economist*, from which an example is shown in Fig. 3.2 [1].

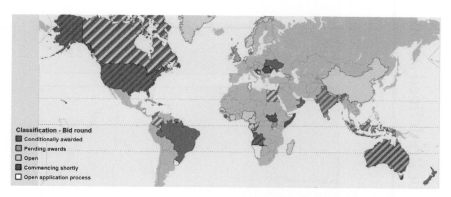

Figure 3.2 Upcoming licensing round overview July 2020 (reproduced with permission from The Petroleum Economist).

The authority representing the oil and gas business within each country will provide further detail of this. For example, the Norwegian Petroleum Directorate (NPD) announces licence awards on a twice-annual basis to encourage participation in the exploration of offshore blocks, as shown in Fig. 3.3. Large numbers of blocks have no exploration activity to date,

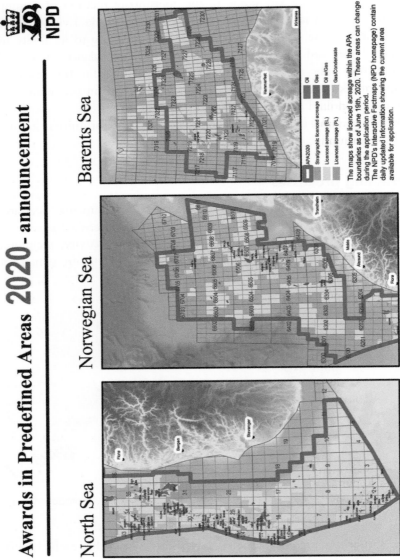

Figure 3.3 Example of offshore licensing awards from Norwegian Petroleum Directorate (NPD).

3.4 Acquisition and disposal

Access to hydrocarbon reserves and resources can be gained by the partial or total acquisition of individual assets or companies. The value of the asset or company to be acquired is based largely on the NPV resulting from future hydrocarbon production. This makes valuation of an oil and gas company different from traditional company valuations, which are based on accounting measures, either the balance sheet value or the profitability of the company. A balance sheet value, or so-called book value, is the balance of the assets and the liabilities of the enterprise. A profit-based valuation is often a multiple of annual profit, or more strictly a multiple of EBITDA (earnings before interest, taxes, depreciation and amortisation). By contrast, the major part of the value of an E&P company lies in the future production, which is still underground, and does not form part of the accountant's estimate of asset value.

The company valuation will require the addition of the value of all its projects that are expected to deliver future production. The final valuation may add some value for fixed assets (plant and machinery) and for goodwill, which could represent a brand value. Nevertheless, the main value lies in the hydrocarbons yet to be produced. Discounted cash flow (DCF) methodology is applied to each project, as described in Chapter 6, and this will require the buyer to make an assumption on commodity prices, which represent a significant uncertainty as described in Section 6.2.

Two other major considerations in the valuation of each project lie in the maturity of the project and the range of uncertainty in the hydrocarbon volumes to be produced. It is helpful to consider these two factors with reference to reserves reporting guidelines.

Many countries (and some companies) have developed their own guidelines for reserves classification, but a common basis, often accepted as the de facto methodology, is the Petroleum Resources Management System (PRMS), published by a collaborative group of the SPE, AAPG, WPC, SPEE and SEG. This was updated in 2018 [6]. The Securities and Exchange Commission (SEC) of the US Government provides an equivalent guideline. Any company listed on the New York Stock Exchange (NYSE) will be required to report reserves according to the SEC guidelines. Countries such as Norway, Canada and Russia have bespoke classification.

The PRMS framework for reserves reporting is shown in Fig. 3.4.

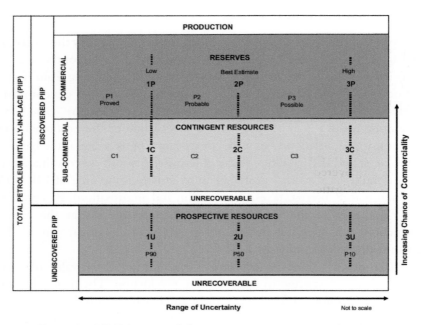

Figure 3.4 PRMS Framework for reporting resources, 2018 Revision.

The principles of resource classification and categorisation depend on the application of three concepts: the **Project** as an activity, classification based on **Project Maturity** and categorisation based on **Range of Uncertainty**.

Under PRMS, each project must be classified individually so that the estimated recoverable sales quantities associated with that project can be correctly assigned to one of the three main classes: **Reserves, Contingent Resources** or **Prospective Resources**. These classes are shown on the y-axis of Fig. 3.4. PRMS breaks each class into sub-classes, but for the purposes of the economic evaluation, the broader definitions are sufficient.

Reserves are associated with projects that the owners can demonstrate to be economic and that have approval for development by all partners. There is no such thing as uneconomic reserves. The reserves class will of course include ongoing projects that are already in production. The economic evaluation of projects in the reserves class follows the straightforward DCF technique explained in Chapter 6.

Contingent resources are associated with projects that are discoveries, but for some reason the project is not yet approved for development. Depending on the sub-class, approval may be contingent upon increased commodity price, agreement of a sales contract, approval from all partners in the project or overcoming some technical hurdle. The contingent resources project will be evaluated as if it were going ahead, to yield an NPV, but then risked by multiplying with a factor called the chance of commerciality (CoC), a number in the range 0.0–1.0. The CoC reflects the opinion of the evaluator that the project will gain approval and move into the reserves class. The risked value ($) is the project NPV ($) multiplied by the CoC (fraction).

Prospective resources are from projects that remain as prospects, i.e. are yet to be discovered. Exploration prospects will fall into this class. While there is no obligation to report Prospective Resources within the PRMS system, if a value is to be attributed to this class of project, then it is evaluated as if it were going ahead, to yield an NPV, but then risked using a CoC, which in this case reflects a combination of risk factors multiplied together. The first factor is the probability of commercial success (P_c) as defined in Section 4.2.3, which is the probability that the produced volumes are sufficient to achieve acceptable economic hurdles (such as $NPV \geq 0$). The second factor, which is occasionally required, is the probability that the prospect can be accessed and developed, for example, if it lies in disputed territory. Again, the risked value ($) is the project NPV ($) multiplied by the CoC (fraction).

On the x-axis of Fig. 3.4 are the categories that represent the range of uncertainty in the volumes to be produced. This range can be determined probabilistically or deterministically as described in Section 4.2.2. If volumes are estimated deterministically, then specific models are generated individually to produce three discrete outcomes. Table 3.2 summarises the nomenclature based on a deterministic approach. Using a probabilistic approach the Low case would correspond to a P90 value, the Best Estimate to a P50 value and the High case to a P10 value.

Table 3.2 Categories of reserves under PRMS using a deterministic approach.

Reserves category	Symbol		Comprised of
Low	1P	P1	Proven
Best Estimate	2P	P1 + P2	Proven plus probable
High	3P	P1 + P2 + P3	Proven plus probable plus possible

Under 2018 PRMS guidelines contingent resources are annotated with the prefix 'C' and prospective resources with the prefix 'U'

This detail becomes important in the evaluation of an acquisition opportunity, since the buyer has to decide on the basis for an offer. Should value be attributed to all three classes (reserves, contingent resources and prospective resources), and which categories should be included (1P, 2P or 3P)? This is of course a matter of choice, but the following provides a typical approach.

If the acquisition target is a single asset such as a discovery with potential for development, then the contingent resources should be evaluated across the whole range of volumetric uncertainty, with the appropriate CoC risk factor applied to the value. The bid is often then based on the risked value of the 1C or 2C case. The same applies to an asset with development approval or a producing asset, with a bid based on the value of the 1P or 2P case. If the acquisition target is a prospective resource, a bid could be based on an estimated risked value of a 1U or 2U case, but this is highly speculative, and some form of risk-sharing deal with the seller is more likely than an offer to purchase the asset.

When the target acquisition is a company owning a portfolio of opportunities in the different classes of project shown in Fig. 3.4, then it is common to add the value of the projects associated with reserves to the risked value of the contingent resources projects, but not to attribute any value to the prospective resources projects. The decision on whether to base the bid on the 1P or 2P categories remains open.

The bidder wishes to make a competitive offer to succeed in acquisition, but not to overpay. A cautious bidder may base the offer on 1P categories, but not win the bid. A more aggressive offer would be based on the 2P categories, but it is rare that the bid is based on the 3P categories. It is worth noting that if the financing for the execution of the subsequent projects is going to be provided by a lender such as a bank, the bank will usually focus on 1P cases only, taking a cautious approach. Any balance between loans arranged with the lender based on 1P development costs and the actual cost of developing a 2P project will need to be met from other sources, such as shareholder capital. Financing options are summarised in Section 3.8.

For senior managers or the Board to make an informed bid, the recommended presentation format for the results of an economic evaluation of a portfolio is shown in Table 3.3. This presents the risked volumes and risked values, but note once again that in deriving a risked value, the un-risked volume is used for the development plan and economics, and the resulting NPV is then risked. The risked volume must not be used for the development plan and economics.

Table 3.3 Summary of portfolio evaluation based on classes and categories of reserves.

Asset	Risked Volume (MMboe)						Risked Value ($m)					
	1P	1C	1U	2P	2C	2U	1P	1C	1U	2P	2C	2U
A	4.0	2.2	0.0	6.5	3.9	0.0	36.0	19.8	0.0	58.5	35.1	0.0
B	2.8	1.5	0.0	4.6	2.7	0.0	16.8	9.2	0.0	27.3	16.4	0.0
C	0.0	3.9	0.0	0.0	5.5	0.0	0.0	42.0	0.0	0.0	65.0	0.0
D	0.0	0.0	5.0	0.0	0.0	7.0	0.0	0.0	35.0	0.0	0.0	50.0
E	3.0	0.0	0.0	5.0	0.0	0.0	30.0	0.0	0.0	50.0	0.0	0.0
Sum	9.8	7.6	5.0	16.1	12.1	7.0	82.8	71.0	35.0	135.8	116.5	50.0
Total		22.4			35.2			188.8			302.3	

The volumes and values should all represent the working interest in the project, which should be the project value multiplied by the working interest. Economic modelling should always be done on a full-field basis, and the results then adjusted for working interest.

Note also that in Table 3.3 the addition of 1P, 1C and 1U numbers for A + B + C + D + E will not equal the probabilistic P90 numbers if the asset value distributions were to be combined probabilistically. The probabilistic combination of five distributions will produce a P90 value that is larger than the arithmetic sum of five individual P90 values. Nevertheless, it is common to arithmetically add 1P values as shown in Table 3.3, albeit giving a conservative result.

When considering the disposal of an asset, exactly the same valuation process should be performed by the seller to establish a fair value. Sellers often publicise the sale with an Information Memorandum summarising the asset, supported by an independent evaluation known as a Competent Person's Report (CPR). This will detail geological information, production forecasts, costs and economics. It is the responsibility of the buyer to review all data available to form their own opinion of the valuation and formulate their bid. Data are made available in a so-called data room, though nowadays this is usually a digital data set that can be accessed remotely.

3.5 Merger

The merger of oil companies is a well-established method of gaining access to E&P opportunities. The industry has a history of major mergers and acquisitions. Examples include BP and Amoco (1998 merger, a $48 bn deal), Exxon and Mobil (1998 merger, an $81 bn deal), Shell acquisition of BG (2016, a $70 bn deal) and more recently Occidental's acquisition of Anadarko (2019, a $57 bn deal).

A merger occurs by mutual agreement at board level for separate companies to join forces and create a new organisation. A company acquisition refers to the take-over of one company by another, which may be by mutual agreement or as a hostile take-over of a distressed company. In this brief section, acquisition will refer to a company acquisition rather than a specific asset as discussed in Sections 3.3 and 3.4.

In either merger or acquisition, the common intent is to gain an improved joint efficiency and to increase access to technology, resources and opportunity. This may be perceived as more effective than increasing the value through organic growth. A common early outcome of a merger is the reduction of staff as a form of removing excess capacity to improve efficiency.

The distinction between acquisition and merger can be somewhat opaque. The merger announced between BP and Amoco in 1998 formed a new entity, BP Amoco, though shortly after the merger the name was changed to BP, placing a question mark over the term merger in this case.

In either a merger or an acquisition, the value of each company must be established to form the basis of the deal, and the asset valuation techniques presented in this book apply for E&P oil and gas field assets at all stages of maturity. The sum of project NPVs will form the majority of value of the upstream sector of a company, to which can be added balance sheet net assets (assets less liabilities), the value of other businesses, such as downstream refining and marketing, and goodwill.

The transaction can be concluded with a cash purchase, a share swap or a mixture of the two. Parts of the agreed payment in the deal may be contingent upon future performance, providing an incentive to deliver future success from projects. Major mergers such as those mentioned are subject to government approvals from its regulators to prevent anti-trust activity, since a monopoly position can be counter to fair competition in the market.

With all good intent, in a merger or acquisition, the integration of business processes and cultures is challenging, and in many cases the results do not meet the expectations at the time of the deal. According to Martin [7], 70%−90% of acquisitions are failures. On that basis, this route to gaining access to E&P projects needs careful consideration and subsequent management of the integration post-merger or acquisition. A plethora of literature provides advice on the elements that do make a success of mergers, and a starting point is from Fubini et al. [8].

3.6 Joint venture and partnership

The terms joint venture and partnership are used rather loosely in the E&P business, but in the context of this section, a JV is a collaboration between companies or entities that join forces to pursue E&P business together. The entities may be private companies or government representatives, examples being the JV of Shell and Esso in the early days of activity in the UKCS, or the JV between Shell and the Omani Government to form a JV entity called Petroleum Development Oman (PDO).

An individual licence block or concession may have a set of partners in the block, each of whom holds an equity share in that block. This partnership may be specific to that particular block, or on occasion, the same partnership can exist in another block. A partner in the block may choose to sell its individual share (or part thereof) in the block at any time, and it is common to have up to 10 partners in the block.

Advantages of both JVs and partnerships include
- the sharing of investment capital, thus reducing the individual's maximum exposure
- sharing of technology and expertise
- combined financial strength that permits access to larger investment opportunities
- access to skills in other market sectors (e.g. downstream, LNG)

However, the drawbacks of these arrangements include
- sharing the project net cash flow
- different investment objectives (long-term (NPV) vs short-term (payback) returns)
- different hurdle rates and targets for project approval
- different economic analysis methods and key assumptions (e.g. commodity price)
- limited access to capital of some partners
- time taken for each partner to agree and approve project proposals

When a JV is formed, it works under the terms of a Joint Operating Agreement (JOA), which is the key commercial contract between the participants in a JV. One member of the JV is nominated as the operator, and is generally responsible for the construction and operation of the wells and facilities, planning and budgeting, and representation of the JV to the host government. The host government may choose (or sometimes insist) to be part of the JV and will be represented by the national oil company (NOC).

Entrepreneurial individuals such as John D. Rockefeller, Henri Deterding, J. Paul Getty, T. Boone Pickens and Armand Hammer founded independent companies that dominated international oil field development for many decades. The dominant private international oil companies (IOCs) of the 1950s were known as the Seven Sisters (Exxon, Mobil, Chevron, Texaco, Gulf, Royal Dutch/Shell and British Petroleum). Daniel Yergin writes a fascinating history of the development of the industry from World War I to the early 1990s in *The Prize* [9].

As petroleum resources became more nationalised, the rise of the NOCs during the 1970—1990s turned the tables on ownership and operatorship of world oil production, with the top 10 IOCs now owning just 5%—7% of the world's proven oil reserves. The top 10 NOCs own some 70%.

Because of this trend, collaboration between independent oil companies and governments has become commonplace in the industry, and many IOCs have chosen to form JVs with NOCs. The advantage to the IOC is access to investment opportunity. The advantage to the NOC is the access to the capital, technical expertise, international exposure and organisational skills that the IOC can bring. Many of the IOCs also have downstream businesses and buyers that can provide market access for upstream production.

Key among the NOCs are ADNOC (Abu Dhabi), CNOOC and CNPC (China), Gazprom and Rosneft (Russia), INOC (Iraq), KPC (Kuwait), NIOC (Iran), NGC (Trinidad and Tobago), NNPC (Nigeria), ONGC (India), PdVSA (Venezuela), Pemex (Mexico), Pertamina (Indonesia), Petronas (Malaysia), Qatar Petroleum (Qatar), Sonatrach (Algeria), Petrobras (Brazil) and Saudi Aramco (Saudi Arabia). The last in this sequence is one of the largest companies in the world, becoming listed for limited private investment in 2019.

3.7 Unitisation and equity determination

Under the terms of a JOA, each party will agree to an equity share in the project. This determines the proportion each party will contribute in terms of funding the project, taking legal and financial responsibility in the contractual arrangement with the host government and sharing the net cash flow or production from the project (known as the lifting rights). The contractual arrangement with the government refers to the system adopted in the host country, which can be a concessionary or contractual agreement as described in Section 6.4.

At points of project investment, cash calls will be made based on each owner's equity share in the JOA. Failure to make a cash call can result in the suspension of a partner's JOA rights in the short term, and potential rejection in the long term for continued non-payment, in which case the remaining partners have the opportunity to take up that share and make the proportional payments.

Agreeing the equity share in a project is made more complicated when the oil or gas field extends across more than one licence block and the ownership in each block is different. To avoid the single field being developed by each licence block owner, which would be inefficient, the field is 'unitised' so that it is developed as a single entity, or unit. A JOA is set up to include all parties with an interest in the field, and one operator is nominated. The proportion of the field in each licence block must be determined prior to development, and this can be done on several different bases, as described in Table 3.4.

Table 3.4 Basis for equity determination in a field straddling licence blocks.

Basis for equity	Requirement	Advantage (+) Disadvantage (−)
Fixed equity for life	A lifetime agreement between the owners of each licence block	+ Simple and quick − May transpire not to represent the proportion of hydrocarbon in place or recovery from each licence block
Areal extent	Map the areal extent of the field to the hydrocarbon water contact	+ Simple and quick − May transpire not to represent the proportion of hydrocarbon in place or recovery from each licence block
Hydrocarbon-bearing gross rock volume	Estimate volume of hydrocarbon-bearing rock in each licence block	+ Simple − May transpire not to represent the proportion of hydrocarbon in place or recovery from each licence block − Does not represent detailed fluid or rock properties

Continued

Table 3.4 Basis for equity determination in a field straddling licence blocks.—cont'd

Basis for equity	Requirement	Advantage (+) Disadvantage (−)
Hydrocarbons in place	Static reservoir model with detailed rock and fluid properties	+ Relatively simple + Accounts for rock and fluid properties − May transpire not to represent the proportion of hydrocarbon in place or recovery from each licence block
Recoverable hydrocarbons	Dynamic reservoir model including development plans across whole producing lifetime	+ Represents recovery from each block − Depends on all future development plans − Costly and time-consuming to build models
Economically recoverable hydrocarbons	Dynamic reservoir model with all development plans and agreed economic criteria for investment and economic cut-offs	+ Represents the value from each licence block − Costly, time-consuming and requires alignment of economic evaluation

The complexity of the initial equity determination increases down the list in Table 3.4. As the field progresses through development and into production, more subsurface data become available, and an initially agreed equity may start to appear incorrect. If the JOA has agreed a fixed equity for life, then there is no recourse to challenge the agreed split — the individuals in the JOA have effectively bought a fixed share of an enterprise, unless they choose to divest their share. On any other basis, however, one set of licence block owners may feel that the initial equity is no longer fair, in which case the JOA will usually allow a redetermination of equity to be called. This invokes a procedure prescribed in the JOA for a review of current data and a series of discussions between licence block owners to agree a revised equity, and rebalancing of entitlements to past revenues and costs. Several equity redeterminations may occur during the lifetime of the field, and these can be time-consuming and costly exercises. The further down the list of Table 3.4 the basis for equity, the more complex and time-consuming this becomes.

The Forties Field in the UK Central North Sea spanned two licence blocks — the majority (approximately 95%) lying in a block owned by BP plus several partners and the remainder (approximately 5%) lying in the adjacent block owned by the Shell/Esso JV. The field started production with BP as operator in 1975. In 1987, an equity redetermination was called, by which time the field had produced over 1.5 billion barrels. The basis for equity, agreed in the JOA, was economically recoverable reserves. To support the revised equity, a full-field static and dynamic reservoir model would be required, along with all plausible future field development and agreed economic cut-off criteria for incremental projects and continued operation. After much discussion, and over a year spent in technical sub-committees, the two parties had agreed a geological model. However, the anticipated time and effort that would be required to build a history-matched full-field dynamic simulation model drove a decision, made at the most senior level within the companies, to close out the redetermination and maintain the current equities for the remaining life of the field. The field remains in production in 2020 under new ownership, as BP divested their share of the asset and operatorship to Apache in 2000.

The lessons learned from that experience are an encouragement to set a simpler basis for the equity determination, and to hesitate before calling a redetermination. When parties are unable to agree on the initial or redetermined equity, the task is often referred to an independent expert, who takes an unbiased position, and follows technical procedures that are specified in the JOA.

3.8 Financing the project

A company can raise finance for E&P projects using generally available financing methods and specialised forms of finance that have been developed for the oil and gas business. The financing options available and the choices made depend upon both the nature of the project and the type of company. In this section, the term project will refer to an activity at any point in the field life cycle presented in Fig. 1.1, which is exploration, field appraisal, development or a further investment during the production phase. This section is a brief summary of financing methods in the E&P business, and more detail can be found in Refs [10—12].

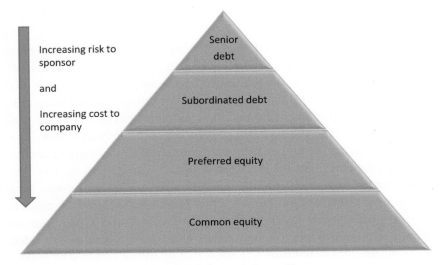

Figure 3.5 Debt seniority ranking.

3.8.1 Debt seniority ranking

In describing options for financing, it is first useful to define the components of the hierarchy in which sponsors of the company call upon the remaining assets if a company falls into liquidation, as outlined in Fig. 3.5. The top of the pyramid represents the sponsors who have the highest priority for repayment. The term sponsors covers the shareholders, who invest their own cash for an equity share of the company, and the lenders who loan money to the company. The loans are termed debt.

3.8.2 Debt financing

Senior debt is usually provided by major banks in various forms of corporate loan (term debt, revolving credit, secured debt). This is the hardest form of finance for small- to mid-sized companies to access since the banks require the company to have a strong balance sheet and secure future revenue from production. Consequently, senior debt financing is best suited to the larger oil companies who own ongoing production. Since senior debt has the first call on repayment, it is generally the cheapest form of financing, but again not available to all companies.

Senior debt can be split into senior secured and senior unsecured debt. Senior secured debt is at the peak of the debt seniority ranking since it is secured against fixed assets as collateral. Only the largest of companies will be able to access unsecured senior debt since it is not backed by collateral, and only guaranteed on the strength of the balance sheet value.

Larger companies can raise capital by issuing bonds, which are also a form of debt. A sponsor buys a bond for a fixed sum, which provides the capital for the company. In return, the sponsor is promised either a fixed interest rate for the term of the bond plus repayment of the original purchase price at the end of the term, or just a fixed sum at the end of the term. That fixed sum will be greater than the original purchase price of the bond. For example, a 10-year bond that pays out $100 at maturity might be issued for $60. This represents an effective interest rate (i) to the sponsor of 5.24%, referred to as the yield of the bond. This can be worked out from $60*(1+i)^{10} = 100$. The lower the cost of the $100, 10-year bond, the higher the yield. The interest rate to the sponsor equals the cost of capital to the company issuing the bond. This is usually lower than the bank loan rate, which is an advantage to the company, and bonds are tradable among sponsors, which adds some flexibility for the sponsor. The sponsor has to have the confidence that the company is robust enough to make good its promise. Secured corporate bonds would be classed as secured senior debt, backed by collateral and therefore high in the debt seniority ranking. Bonds may also be unsecured.

Subordinated debt, also known as junior debt, is typically provided by smaller banks, or by the parent company to a subsidiary, or by the shareholders themselves. When lending to the company, sponsors will scrutinise the strength of the company's balance sheet (looking at asset value) and the track record of the profit and loss account (looking at profitability), and these should be presented in good order by the company seeking the loan. The type of project that the company intends to invest in will also be reviewed, and often the company will commission a CPR from an independent party to present to the potential lender.

3.8.3 Equity financing

Issuing equity to raise finance is an option for all sizes of company. When pursuing a major investment or acquisition, IOCs may choose to issue new shares in the company in return for cash. This dilutes the percentage ownership of the original shareholders who will then have a smaller share in a company with a larger balance sheet asset value, since the cash raised from the share issue will either become part of the current assets (as cash) or fixed assets (the value of the equipment invested in). The original shareholders are thus neutral to the new share issue.

Issued shares can be classed as preference shares or common (ordinary) shares, as illustrated in Fig. 3.5. Preference shareholders typically do not receive voting rights, but unlike ordinary shareholders may have an agreed level of dividend payment, made at specified points in the calendar year. In the case of a company liquidation, preference shareholders are ahead of ordinary shareholders for payment, and thus take lower risk. The preference shareholder dividend may actually be lower, but more assured, which appeals to certain investors.

Shares can be offered to a wide audience including the general public, and private equity groups. It should be noted that the private equity investor group may choose to take an active role in the management of the company, perhaps on the technical front, but almost certainly on the financial side, which can add expertise, or be seen as intrusive by the core management team.

Since both forms of shareholder are taking higher risk than the sponsors who provide debt, the cost of equity is higher than the cost of debt.

Equity finance is an option for small- to medium-sized companies looking to undertake exploration and appraisal, by offering existing shareholders the opportunity to invest further cash in return for additional shares, or inviting third parties to take up a shareholding.

However, when a development project requires investment, the small- to medium-sized companies who do not yet have revenue from a production stream may not be able to access senior debt or raise enough equity finance, and some intermediate form of lending is required. In these circumstances, mezzanine financing may provide a solution.

3.8.4 Mezzanine financing

The company seeking mezzanine loans is likely to have a modest balance sheet and profitability, no significant existing production or proven reserves, and therefore cannot yet access senior debt. For such smaller- to medium-sized companies, with projects in the early development phase of the life cycle, mezzanine loans provide a potential stepping-stone between equity and senior debt.

As the name suggest, mezzanine loans are a hybrid form of debt, with a risk level between equity and senior debt. The providers of mezzanine loans are often private equity groups or consortia of smaller banks, rather than the larger banks. Due to the limited collateral of the company, the sponsor is taking higher risk and therefore increases the loan interest rate compared to the cost of senior debt, often making mezzanine loans an expensive option.

Conditions often apply to the loan, such as the sponsor monitoring of investment and operations, and setting covenants based on the financial progress of the company. If the target ratios in the covenants are broken, harsher terms for the loan may be invoked. Another potential condition of the loan is that the sponsor takes some equity share in the production, once achieved. This may take the form of a fixed overriding royalty on production or gross revenue, or a shared ownership of the assets.

Mezzanine financing originated in the United States, and has extended globally. It offers a stepping-stone to senior debt or to reserves-based lending (RBL) (see Section 3.8.5), but can be an expensive and restrictive solution for the small- to medium-sized company trying to get a project into production. Good examples of the use of mezzanine and other forms or financing E&P projects are given in Ref. [10].

3.8.5 Reserves-based lending

RBL is a financing mechanism specific to the E&P industry. This becomes an option once the company can demonstrate that it owns hydrocarbon reserves. Reserves will only become valid when the company has some ownership of a project with a development plan and adequate development approvals and partner agreements, as defined in the guidelines used for reserves and resources classification discussed in Section 3.4. Contingent and prospective resources do not qualify for RBL.

Loans based on RBL can be in the form of a general credit facility that provides flexibility as to when the company draws down the money, or a simple term loan. The lender is often a bank or a consortium, and requires detail of the project or collection of projects against which the loan is provided. The loan is based on the NPV of the company's share of the projects. The detail of the projects must therefore be sufficient for the lender to calculate the project NPV, using the methods described in Chapter 6, and will include production profiles, capex and opex and the fiscal terms. The valuation of the assets may be referred to an independent consultancy to produce a CPR, using a price forecast provided by the lender or an independent view such as the Wood Mackenzie oil price forecast referred to in Section 6.2.

The profiles used for the NPV calculation are based on 1P reserves, as defined in Table 3.2, if using the deterministic definition of proven reserves. This category may be further broken down into Proved Developed Producing (PDP), Proved Developed Non-Producing (PDNP), and Proved Undeveloped (PUD), with different weighting on each sub-category, say 100%, 75%, 50%. This is common in the United States. Probable (P2) and Possible (P3) reserves do not form part of the contribution for RBL.

The lender is taking a relatively low technical risk by basing the loan on the NPV of 1P reserves, since the proven reserves should have at least a 90% probability of being delivered. Price risk is a more significant risk factor, and there are likely to be clauses in the loan agreement that trigger revisions if oil prices vary significantly from the base assumptions. In addition, the agreement will contain covenants relating to measures such as the company's current assets, debt levels and capital expenditure. If covenants are broken, then the terms of the loan may be adjusted.

The RBL option is more popular in the United States and Western Europe where political risk may be considered lower, and is better suited to small- to mid-sized companies. With the constraints that RBL imposes, oil majors tend to prefer issuing shares and bonds or taking corporate loans. With a strong corporate balance sheet, the oil majors can generally borrow at a lower cost of capital, and avoid borrowing against specific projects, which is referred to as project financing.

3.8.6 Volumetric production payments

This mechanism for raising finance exists primarily in the United States, with its origins in the mining industry. In a volumetric production payment (VPP) arrangement, the E&P company secures funds from a sponsor prior to production commencing, and then transfers a portion of the production to the sponsor over time. The pre-payment allows the E&P company to fund the development and production of an asset, or a group of assets.

The sponsor takes the commercial risk on the prevailing oil price, and may choose to hedge the sale of oil or gas on the futures market. To limit the technical risk, the sponsor will require a CPR or equivalent, and the agreement will be based on the proven reserves, or some part thereof, as discussed in Section 3.8.5. In addition, the sponsor may reduce the technical risk by basing the up-front payment on production from a number of fields, rather than a single development.

3.8.7 Islamic finance

Specialised forms of finance have been developed to comply with Sharia principles or Islamic Law, which prohibit the earning of fixed interest for lending money. This is applicable in many parts of Asia and Africa, and to most traditional members of OPEC.

The sponsor makes a capital contribution to the project (the equivalent to a monetary loan) based on the value of the fixed assets in the project, so the sponsor is actively participating in a project. Instead of charging interest on a loan, the sponsor takes a share of the profit from the project. The sponsor is taking a direct risk on the project delivering its forecast profit level, and as for RBL and mezzanine finance, the proven reserves and development plans will be scrutinized.

An alternative form of finance, which also complies with the Islamic principles, is a sale-and-buy-back contract. The sponsor is sold the asset that the E&P company wishes to develop for a sum of money, which is the equivalent of a loan. The E&P company then buys the asset back either as a normal purchase or through a series of lease payments. The margin of profit between the sale price (paid by the sponsor) and the buy-back price (paid by the E&P company) replaces conventional interest.

Islamic finance is expected to become more widespread as the balance of global hydrocarbon production and general commerce shifts to the Islamic world, parts of which have built up a high level of liquidity in sovereign wealth funds and has ample capacity to provide finance.

In summary to Section 3.8, there are many financing options available, including hybrids of those mentioned in this section. The appropriate form of finance will be determined by the size of the company and the stage at which the project lies within the field life cycle.

References

[1] The Petroleum Economist. https://www.petroleum-economist.com/licensing-rounds.
[2] UK Oil and Gas Authority. https://data-ogauthority.opendata.arcgis.com/datasets/.
[3] The Economist Intelligence Unit. https://www.eiu.com/landing/risk_analysis.
[4] Refinitiv. https://www.refinitiv.com/en/products/country-risk-ranking/.
[5] C. Rice, A. Zegart, Managing 21st-Century Political Risk, Harvard Business Review, May—June 2018 Issue. https://hbr.org/2018/05/managing-21st-century-political-risk.
[6] Petroleum Resources Management System (Revised June 2018), SPE/WPC/AAPG/SPEE/SEG/SPWLA/EAGE, ISBN 978-1-61399-660-7.
[7] R.L. Martin, M&A: The One Thing You Need to Get Right, Harvard Business Review, June 2016.
[8] D. Fubini, C. Price, M. Zollo, Mergers: Leadership, Performance and Corporate Health, Palgrave Macmillan, 2006, ISBN 978-0-230-80075-5.
[9] D. Yergin, The Prize, Simon & Schuster Ltd, 1991, ISBN 0-671-71189-X.
[10] Duff & Phelps, Financing Instruments in the Upstream Sector, Oil & Gas Intelligence Report, April 2018. https://www.duffandphelps.com/-/media/assets/pdfs/publications/valuation/oil-and-gas-intelligence-report-upstream-sector-2018.ashx.
[11] S. Szczetnikowicz, J. Dewar, Practical Law Practice, Financing Options in the Oil and Gas Industry, Thomson Reuters, 2018. https://www.milbank.com/images/content/9/7/v2/97930/Financing-options-in-the-oil-and-gas-industry.pdf.
[12] R. Clews, Project Finance for the International Petroleum Industry, Elsevier, 2016, ISBN 978-0-12800-158-5.

CHAPTER 4

Exploration

Contents

4.1 Introduction	69
4.2 Risk and uncertainty in exploration	70
4.2.1 Prospect risking	72
4.2.2 Volumetric uncertainty	75
4.2.3 Combining exploration risk and volumetric uncertainty	83
4.2.4 The risk-reward balance	88
4.2.5 Decision tree analysis — an introduction	89
4.2.6 Expected monetary value in exploration using decision trees	91
4.2.7 Summarising the opportunity	94
4.2.8 Formulating the bid — theory and practice	96
References	100

4.1 Introduction

Assuming that the investor has set up the right conditions to make a bid to explore in a region, the next step is to formulate the bid for submission in the licencing round. The objective will be to win the licence award without over-bidding, and this chapter will introduce techniques for formulating an appropriate and defensible bid.

This stage of the field life cycle is where oil and gas companies are seen to truly compete — explorationists will argue that obtaining the right acreage is the key to business success, and that subsequent field development is a matter of applying the appropriate technology, which is much more generally shared across the industry than intelligence on exploration prospects.

Each bidder will formulate their own view of the value of a licence, and the individual bids submitted will invariably be different. Reviews of bid costs against value of outcomes have demonstrated that in general oil and gas companies over-bid (Capen et al.) [1]. Megill [2] indicates that in offshore licence bids, the average winning bid is significantly higher than the second bid, by as much as a factor of two. This may be a result of the log-normality of the volumetric estimates, explained in this chapter, or the enthusiasm to win the bid. With such potential to leave money on the table, it is important to develop a sound method of formulating the bid value that demonstrates that the investors are not spending more than the technically justifiable value. Otherwise, there must be a transparent over-riding reason to make an aggressive bid, sometimes justified as a strategic choice.

4.2 Risk and uncertainty in exploration

In the context of exploration, uncertainty and risk are illustrated in Fig. 4.1 as two halves of the same challenge to estimate the potential hydrocarbon volume.

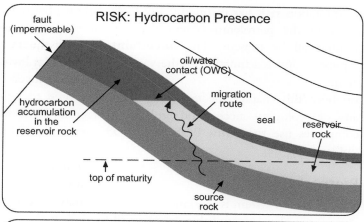
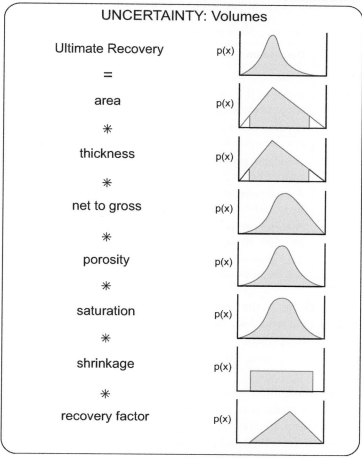

Figure 4.1 Risk and uncertainty in the exploration context.

The exploration **risk** lies in finding any hydrocarbons, for which the components of the petroleum system depicted in the upper half of Fig. 4.1 must all be present. The **uncertainty** lies in the range of volumes present, and is a function of the variables listed in the lower half of the figure.

Throughout this chapter the terms P_g and P_c are used to represent probability of geological success and probability of commercial success. Some companies use the terms POS_g and POS_c or COS_g and COS_c (chance of success).

4.2.1 Prospect risking

When the petroleum system is working favourably to form a hydrocarbon accumulation, the elements in Table 4.1 must all be present.

Table 4.1 Elements of the petroleum system.

Element	Description
Source rock	An organic-rich sediment
Maturation	Burial of the source rock to sufficient pressure and temperature to generate oil or gas
Migration pathway	A permeable pathway by which the hydrocarbons can migrate towards a trap
Trap	A suitable geometric shape in which hydrocarbons can accumulate (e.g. an anticline, fault trap, stratigraphic trap)
Seal	A non-permeable formation preventing escape of hydrocarbons from the trap (e.g. shale, salt)
Reservoir	A rock with porosity (e.g. sandstone, carbonate) that provides storage space for the hydrocarbons
Timing	The reservoir, trap and seal must be present prior to migration occurring

Using knowledge of the basin history, analogue information and seismic, gravimetric or magnetic data, the geophysicist and geologist will estimate a probability of the presence of each of these elements. This will be a number between 0.0 and 1.0 (0 being not present and 1.0 certain to be present).

For an accumulation to occur, all of these elements must be present. The equation of joint probability in statistics states that the joint probability that two events (A and B) will occur together P(AB) is the product of the individual probabilities P(A) and P(B), assuming the events to be independent.

$$P(AB) = P(A \text{ and } B) = P(A) \times P(B)$$

The elements of the petroleum system are generally assumed independent, and so applying this rule to estimate the probability of geological success (P_g) becomes

$$P_g = P(\text{source}) \times P(\text{maturation}) \times P(\text{migration}) \times P(\text{trap}) \times P(\text{seal})$$
$$\times P(\text{reservoir}) \times P(\text{timing})$$

P_g refers to the probability of making a discovery of some hydrocarbons, without specifying the volume discovered. As an explorationist, proving that the petroleum system works may be considered a success. The engineering and commercial teams will also be considering whether the volume discovered is sufficient to justify an economic success, and this will lead to the estimate of the probability of commercial success (P_c), covered later in this chapter.

As a simple illustration, if each of the elements of the petroleum system was estimated to be 0.5, the P_g would be 0.5^7 (i.e. 0.0078, 0.78%, or 1/128). The probability of success reduces alarmingly quickly due to the equation of joint probability, and this introduces the different approaches that may be taken to assessing the P_g.

At one extreme of the approaches, each element is assessed individually, and further split into sub-classes of *presence* and *effectiveness*. For example, a reservoir may be *present* and yet not good enough quality in terms of flow rate (dictated by the permeability of the rock and the viscosity of the fluid) to be *effective* as a commercial producer. In this approach,

$$P(\text{reservoir}) = P(\text{reservoir presence}) \times P(\text{reservoir effectiveness})$$

This approach can also be applied to the seal, trap and migration pathway, leading to many more inputs required in the estimation of Pg. At the extreme, by splitting the estimate into *Play Chance* and *Prospect Specific Chance*, there can be up to 32 elements to the P_g estimate.

The approach of risking an exploration opportunity by combining its Play Chance and Prospect Specific Chance is covered well by Allen and Allen [3] and may be considered as a matter of scale.

On a broad scale (10–100 km) the geoscientists consider the evidence to estimate the presence of mature source rock, sealing rock and reservoir rock. Focussing in on a specific area of interest (1–10 km), sparked often by seismic responses, the geologist then considers an identified prospect in terms of trap, seal, reservoir, migration and timing.

Table 4.2 shows how these two estimates would be combined, in this case resulting in a P_g estimate of 16%, approximately 1 in 6 chance of geological success:

Table 4.2 Example of Play and Prospect Specific risking for P_g.

PLAY CHANCE (%)	Presence	Effectiveness	Maturity	Migration	Chance	Comment
Reservoir	60	90			54	On the edge of play
Seal	90	90			81	Possibly off sand fairway
Source	100		100	100	100	Proven by nearby field
					44	PLAY CHANCE (%)
PROSPECT SPECIFIC (%)	Presence	Effectiveness	Maturity	Migration	Chance	
Trap	100	100			100	Good structure on seismic
Seal	100	90			90	Risk of breach at top seal
Reservoir	90	90			81	Not specific to prospect
Charge	100		100	50	50	Remote from known charge
					36	PROSPECT CHANCE (%)
					16	POSG (%)

In Table 4.2, the term Charge has been used, to combine source, timing and migration, and is an example of grouping or lumping elements together rather than splitting them into individual components.

Therefore, at the other extreme of approaches, the explorationist may choose to group the elements together into a limited number of considerations (Table 4.3).

Table 4.3 Grouped terms for estimation of P_g.

Grouped term	Includes
Source	Source rock, maturity and migration
Trap	Trap and seal presence and effectiveness
Reservoir	Reservoir presence and effectiveness

The danger of splitting elements rather than lumping is that of over-risking the prospect. Human bias makes it difficult to assign a probability of 1.0 to an element, as this implies certainty. We often tend to be cautious in our estimates. But since the probability of each element is multiplied together, note that the result of multiplying seven inputs each with a 0.9 probability of occurrence becomes $0.9^7 = 0.48$. Each element is quite certain but the product becomes significantly uncertain.

If there is clear evidence within the basin containing a prospect that one element is present, such as source rock, then applying a P(source) = 1.0 would be reasonable. If there is doubt about the migration pathway feeding into a trap then the detailed consideration should be given to P(migration). In this case splitting out the elements clarifies where the uncertainty lies.

The key to risking exploration opportunities is to provide a consistency of approach so that each one receives the same treatment. Consistency may be achieved by having an agreed risking procedure, laid out in company guidelines, and presenting results to a risking committee or a peer review team.

4.2.2 Volumetric uncertainty

Assuming hydrocarbons have been discovered, the second half of the explorationist's challenge in supporting the estimate of the exploration prospect value is capturing the range of uncertainty in the volume that could be recovered. Fig. 4.2 shows the input variables involved in making an estimate of the recoverable volumes and the measurements on which they are commonly based.

Variable	Symbol [unit]	Basis	Description	Image
Area	A [m²]	Mapping based on seismic data	Areal extent of accumulation to hydrocarbon-water contact	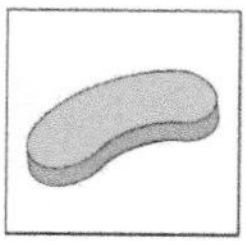
Thickness	T [m]	Seismic imaging or well logs	Gross thickness of hydrocarbon bearing reservoir	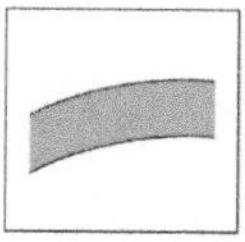
Net-to-gross	NTG [fraction or %]	Well logs or analogue	Ratio of net reservoir to gross thickness	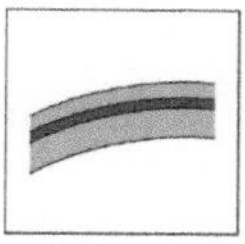
Porosity	Ø [fraction or %]	Well logs or analogue	Porosity of the reservoir units	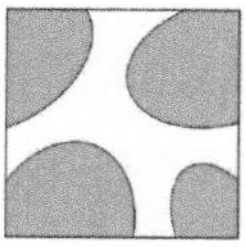
Hydrocarbon Saturation	S_h [fraction or %]	Well logs or analogue	Hydrocarbon saturation within pores	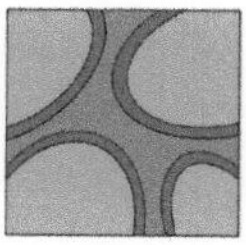
Shrinkage	1/Bo (oil) or 1/Bg (gas) [st m3 / reservoir m3]	Fluid analysis or analogue	Fluid volume at reservoir conditions per volume at standard conditions	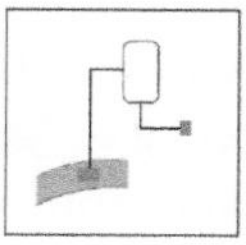
Recovery Factor	RF [fraction or %]	Development plan or analogue	Fraction of hydrocarbons in place to be recovered through development	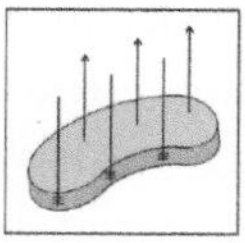

Figure 4.2 Input variables to estimate recoverable hydrocarbon volumes.

Note that for the shrinkage factor, the terms B_o and B_g are introduced for oil and gas fields. B_o is known as the oil formation volume factor, and B_g the gas formation volume factor, and will typically be estimated by the reservoir engineer based on fluid sample analysis.

The following terminology is useful, and the common abbreviations are adopted in this book.

STOIIP : stock tank oil initially in place

$$\mathbf{STOIIP} = A \times T \times NTG \times \varnothing \times S_o \times 1/B_o [m^3]$$

GIIP : gas initially in place

$$\mathbf{GIIP} = A \times T \times NTG \times \varnothing \times S_g \times 1/B_g [m^3]$$

UR : ultimate technical recovery

$$\mathbf{UR} = STOIIP \times RF \quad \text{or} \quad GIIP \times RF \quad [m^3]$$

EUR : economic ultimate recovery

$$\mathbf{EUR} = UR \text{ but truncated by applying a commercial cut-off} [m^3]$$

Remaining reserves

$$\mathbf{Remaining\ reserves} = EUR - \text{cumulative production} [m^3]$$

To classify as reserves, recoverable volumes must be both technically and commercially viable. So, to avoid confusion this book will refer to recoverable volumes or ultimate recovery, implying that no commercial cut-off has yet been applied. Chapter 6 will further discuss commercial cut-offs to production profiles.

Care must be taken with units — the values above have used metric units, though 'field units' are also common within the oil industry.

It is easy to misuse the formation volume factor supplied by the reservoir engineer, as the shrinkage is $1/B_o$ or $1/B_g$. It is useful to remember that when an oil volume is brought from reservoir conditions to stock tank conditions (usually 60°F and 1.0 atm) it shrinks as it releases dissolved gas. Free gas (gas in a separate phase), on the other hand, expands when taken from reservoir to standard conditions.

When combining the input variables for an HCIIP (hydrocarbon initially in place, a general term covering both oil and gas) the subsurface team estimates a range of values for each variable. At the exploration stage, this is dominated by the seismic interpreter and geologist (A, T), supported by the petrophysicist (NTG, Ø, S_h) and reservoir engineer (B_o, B_g, RF).

The range of uncertainty in recoverable volume estimation can be made in a number of ways, including
- **Deterministic** methods — discrete models generated using specific assumptions to create several possible realisations
- **Geostatistical** methods — modelling variability of variables using mathematical methods to create multiple possible realisations
- **Probabilistic** methods — modelling input variables with continuous ranges of values to generate a continuous range of possible recoverable volumes, and associated probabilities

Ringrose and Bentley [4] and the SPE Petroleum Resource Management System Guidelines [5] detail these alternative methods of capturing uncertainty in reservoir volumes and will be referenced further in Chapter 5.

At the exploration stage, it is most common to use the probabilistic approach, as the limited data available are often insufficient to justify

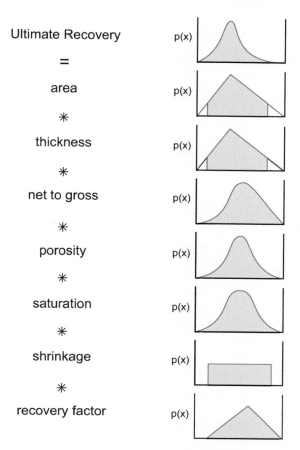

Figure 4.3 Distribution of input variables to estimate ultimate recovery (UR).

building discrete models. The ranges of uncertainty can be represented by a probability density function, illustrated in Fig. 4.3.

A continuous range of possible values, with associated probabilities, represents each input variable. The probability density functions used earlier are illustrative only, and are not necessarily the recommended distribution type for each variable. The selection of the distribution type will be discussed in more detail in Chapter 5. Fig. 4.4 summarises the most common distribution types, and the way in which they are represented using simple parameters such as minimum, most likely, maximum, mean (μ) and standard deviation (σ or s.d.).

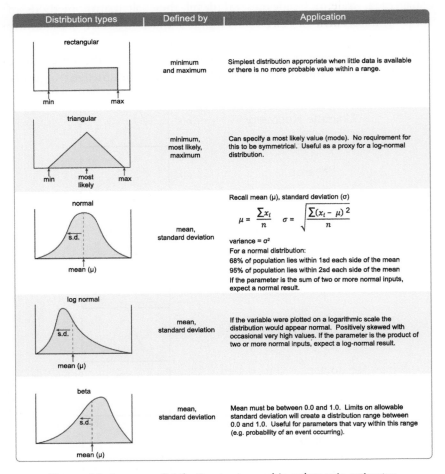

Figure 4.4 Common distribution types used in volumetric estimates.

In Fig. 4.4, the x-axis represents the value of the variable, and the y-axis, the associated probability of occurrence.

The benefit of parameterising the distributions is that the range can be represented by a limited set of numbers. For example, the standard deviation (σ) and mean (μ) defines the whole of a normal distribution. In addition, distributions can be combined together using simple rules or statistical methods.

The variance is the square of the standard deviation (variance $= \sigma^2$). This becomes useful for combining distributions.

The log-normal distribution justifies some further explanation, since a recoverable volume estimate is inevitably log-normally distributed. A log-normal distribution can be parameterised by the mean and standard deviation, and would appear normal if the x-axis was plotted on a logarithmic scale. For a normal distribution, the mean, mode (most commonly occurring value) and median (the value which splits the ranked values 50/50 higher and lower, also known as the P50) are the same. In a log-normal distribution, these values are not coincident, as Fig. 4.5 illustrates.

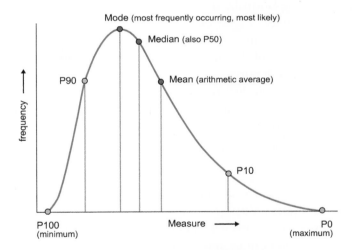

Figure 4.5 Log-normal distribution values.

The mode does not carry particular significance statistically, but note that the mean (arithmetic average) exceeds the median (P50), since the mean is influenced by the occurrence of occasional very high values in the distribution. The P90 value implies that 90% of the population is equal to or greater than that value, and the P10 value implies that 10% of the population is equal to or greater than that value.

The reason that the estimate of recoverable volume is typically log-normally distributed lies in a statistical property of multiplication of input variables. When input variables are multiplied together the result tends towards a log-normal distribution, but if added together the result tends towards a normal distribution, as illustrated in Fig. 4.6.

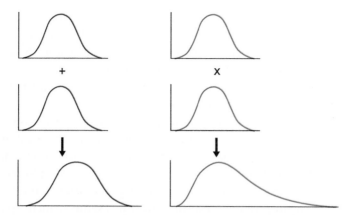

Figure 4.6 Addition and multiplication of input variables.

Handy statistical rules may be applied to the distributions when inputs are normally or log-normally distributed.

Addition of inputs:
- The mean of the sum is the sum of the means
- The variance (σ^2) of the sum is the sum of the variances

Multiplication of inputs:
- The mean of the product is the product of the means
- The coefficient of variation (σ/μ or CV) of the product can be calculated as follows:

$$\left(1 + CV_{product}^2\right) = \left(1 + CV_a^2\right) \times \left(1 + CV_b^2\right) \times \left(1 + CV_c^2\right) \text{ for inputs a, b, c}$$

noting that the coefficient of variation is defined as the standard deviation divided by the mean (σ/μ).

So, as long as input variables can be described as normal or log normal, these parametric rules can be applied to estimate the result of either adding or multiplying inputs. This is the basis of the 'parametric method' of estimating results.

This becomes an unnecessary method if one has access to a simple tool to apply Monte Carlo simulation, a technique that was developed during World War II to predict the outcome of combined inputs (for an interesting history of this, see Ref. [6]).

Referring back to Fig. 4.3, a reasonable estimate of the mean of the recoverable volume could be made by multiplying together the mean values of the input variables. The parametric rules above would support this. For a single estimate, this would be defensible.

The temptation to multiply together low cases to obtain a low case estimate would, however, yield an extreme low result and would imply that all input variables were simultaneously low — an extremely low probability. Multiplying together P90 input values will not yield a P90 result — in fact multiplying three P90s together will give approximately a P98 result.

The Monte Carlo technique randomly samples the input values, assuming in the simple case that input variables are independent, i.e. occurrence of a low value of one variable does not imply a low value of another. In a single iteration, a random number generator is used to sample each input variable individually and then combines the values of each variable according to the relationship defined by the user. This produces a single result. The process is repeated over many iterations to produce many values of the result, which can then be plotted as a probability density function, as illustrated in Fig. 4.7.

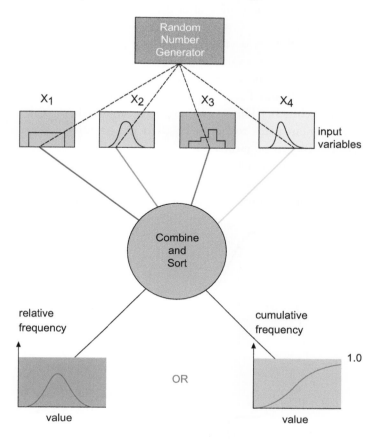

Figure 4.7 Monte Carlo simulation schematic.

Any value and its corresponding cumulative probability may be extracted from the resulting distribution. This becomes particularly straightforward if the output is plotted as a cumulative probability curve as shown on the bottom right. It is recommended that a large number of iterations is run (10,000 will generally provide a smooth and reliable result). Commercial software packages such as Crystal Ball [7] or @RISK [8] or shareware MCApp [9] may be used to perform this task. The commercial packages can handle dependency between input variables when this is appropriate, by specifying a correlation coefficient.

Once a continuous distribution for the recoverable volume has been generated using the Monte Carlo simulation technique, it is common practice to take a limited selection of resource volumes and establish the value of a field development in these cases. This requires an estimate of the gross revenues from the project (production multiplied by price), the technical cost of development (capex, opex), and fiscal costs (taxes, royalty) to yield the resulting NPV after discounting the net cash flow, as described in Chapter 1.

It is a relatively arbitrary decision as to which points on the continuous distribution to select for evaluation. Any number of points on the continuous distribution could be taken — some companies choose five (P90, P70, P50, P30 and P10), but most commonly the P90, P50 and P10 values are chosen. This is a pragmatic approach, which allows discrete development plans to be attached to the resource volumes, and hence discrete NPVs to be estimated.

When taking the next step in the overall evaluation of an exploration prospect, it becomes necessary to attach relative weightings to each case selected, and for this Swanson's approximation is useful (Fig. 4.8).

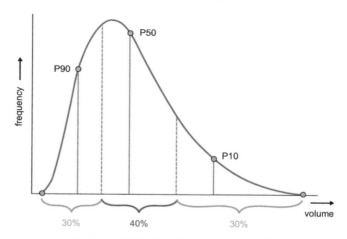

Figure 4.8 Swanson's approximation to weighting discrete values.

In Swanson's approximation, the P90 is the mean of some fraction of the population. By definition, 10% of the population is below the P90 value, so a graphical method is used to determine what fraction of the population must be above the P90 such that the P90 becomes the mean. This is approximately 20%, making the P90 the average of approximately 30% of the population. The same reasoning is applied to the P10, which is the mean of approximately 30% of the population. The P50 thus represents the mean of the remaining 40%. So, when using P90, P50, P10 values from the distribution, they should be weighted 30%, 40%, 30%. This is an approximation, less true for highly log-normal distributions.

When using Swanson's mean, rather than the statistically generated mean, this is defined as

$$\text{Swanson's mean} = 0.3 \times (P90) + 0.4 \times (P50) + 0.3(P10)$$

Further discussion can be found in Ref. [10].

4.2.3 Combining exploration risk and volumetric uncertainty

The previous sections in this chapter described exploration risk as the probability of success of finding hydrocarbons (P_g) and uncertainty in exploration as the range of resource volumes, should hydrocarbons be present. These must now be combined to estimate the value of an exploration opportunity so that a bid can be made on a licence or the cost of an exploration programme can be justified.

The range of volumetric uncertainty has been presented as a probability density function, or frequency diagram. Alternative presentations are shown in Fig. 4.9.

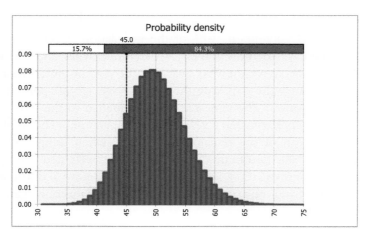
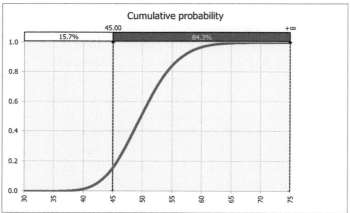
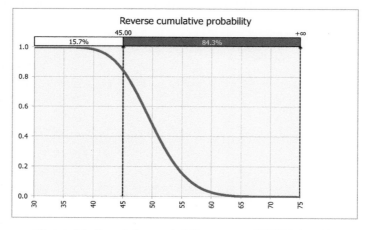

Figure 4.9 Cumulative probability curves of STOIIP (x-axis).

The results shown in Fig. 4.9 are generated for a variable with a mean of 50 and standard deviation of 5, using @RISK. This format is useful in two ways.

Firstly, it is straightforward to extract a value for a particular percentile, e.g. the P90, P50 or P10 value. In the example above, from the cumulative probability curve, it is straightforward to read from the y-axis that there is a 15.7% chance that the value is less than 45.

From the reverse cumulative probability curve, there is an 84.3% chance that the value is greater than 45. The reverse cumulative probability density function may also be referred to as a 'probability of exceedance curve' or an 'expectation curve'. As a reminder, for a P90 value there is a 90% probability that the variable exceeds that value.

Secondly, the reverse cumulative format can be used to conveniently combine the range of volumetric uncertainty with the P_g, as shown in Fig. 4.10.

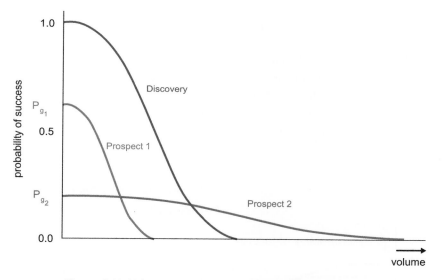

Figure 4.10 Volumetric uncertainty and exploration risk (P_g).

In this example, the discovery has a 100% probability of presence of hydrocarbons. Prospect 1 may be described as a low-risk, low-reward opportunity — a P_g of 0.65 would be considered as a very high chance of success, perhaps appropriate in a well-established basin or a near-field prospect, close to existing discoveries. Prospect 2 may be viewed as high-risk, high-reward, with a P_g of 0.2 (1 in 5) but the potential for large volumes. Which is the more attractive of the prospects is a matter of taste for the investor. Prospect 1 is cautious while Prospect 2 is more aggressive. This choice can only be made by further analysis of the technical complexity and costs of development, the NPV of the projects and understanding of the profile and attitude to risk of the investor. In all cases, the probability of failure is $(1-P_g)$, incurring a loss equal to the cost of the exploration activity.

A useful next step in evaluating such opportunities is to estimate the capex and opex of development, the revenues from a production profile, the government take and then run preliminary economics to establish the minimum commercial field size (MCFS) that would yield a break-even NPV (i.e. NPV = $0 when discounting at the investor's cost of capital). This MCFS can be used to estimate the Probability of Commercial Success (P_c), also known as the Probability of Economic Success (P_e).

For example, consider exploration where basin modelling predicts finding gas rather than oil, but no local infrastructure exists to allow immediate local use of gas. The costs of development would include not only the cost of the gas field development but also the development of a local infrastructure (gas to power plant, or gas distribution network) or possibly an LNG (liquefied natural gas) plant and export facility such as a loading jetty. The MCFS required in this case may be in the order of 10–20 trillion standard cubic feet (Tcf) of recoverable gas. MCFS may also be termed Minimum Economic Field Size (MEFS). Singh et al. [11] provide detailed methodology for establishing MEFS.

The MCFS is applied to the reverse cumulative probability curve shown in Fig. 4.11. The P_c (also known as P_e, probability of economic success) is always less than the P_g.

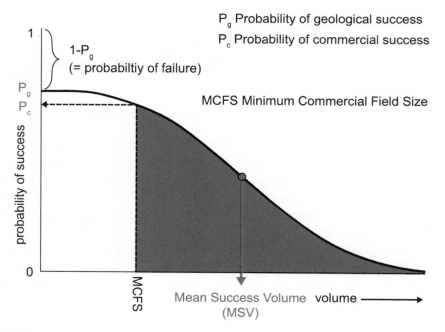

Figure 4.11 Minimum commercial field size (MCFS) and probability of commercial success (P_c).

In using PrecisionTree® [13], the format of this example would be as shown in Fig. 4.15.

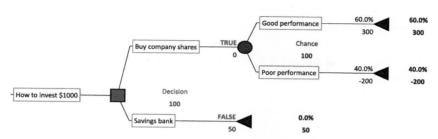

Figure 4.15 Simple decision tree in PrecisionTree® format.

The user inputs the black figures. At a chance node, the branches contain the probabilities on the top of the branch and the net payoffs below. Any cost of buying the shares (a broker's fee) would be captured below the decision to buy shares (zero in this example). The blue and green figures are calculated. The blue numbers at the terminal nodes are the net payoffs and the probability of realising that outcome — in this example there is zero probability of making $50 through investing in the bank, since that is not the optimum choice. The software annotates the preferred decision as 'TRUE' and the rejected option as 'FALSE', noting the EMV of the better decision in green a 'DECISION' with a value of $100. Of course, if the net payoffs or the probabilities were to change, the decision may switch — this could be investigated using sensitivity analysis on either of these inputs.

4.2.6 Expected monetary value in exploration using decision trees

The EMV of an exploration prospect may be approached using DTA or by a formula, but DTA is recommended as it helps to clarify net payoffs and provides a visual image of the timeline that can invite comment and help to clarify the possible outcomes of exploration.

The main decision in exploration, at the root of the tree, is how much to spend on exploration of an opportunity. DTA may indicate spend nothing and walk away, or may justify a significant investment. The exploration spend will be a combination of study work, technical activity (surveys, drilling), signature bonus payable on success in a licence round and any forfeits due on releasing a block. Estimating the EMV of an undrilled prospect or licence will guide the level of commitment, in particular a bid value for a licence.

Several approaches may be taken to the evaluation, annotated as Cases 1, 2, 3:
1. Use the P_c and only evaluate the MSV case
2. Use the P_c and assume a range of outcomes which exceed the MCFS would be developed
3. Use the P_g and assume a range of outcomes, each of which would be developed

Figs 4.16 and 4.17 illustrate the approaches to estimating the EMV of an oil prospect, and in each case the exploration cost is represented as E.

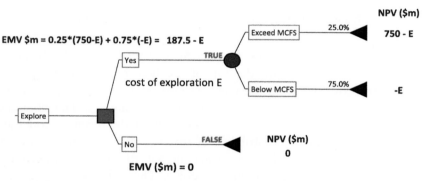

Figure 4.16 EMV based on the MSV development only (values in $m).

Case 1, shown in Fig. 4.16, assumes that the P_c is 0.25 and the NPV of developing the MSV is $750m. The NPV of the development project does not include any cost of exploration, so the values at the terminal nodes following exploration must carry the cost of exploration, E in this example. The EMV calculated then includes the cost of exploration, technical costs of exploration and related costs such as a signature bonus. This is the simplest approach since only one project cash flow is evaluated — that of developing the MSV.

Case 2, shown in Fig. 4.17, evaluates three cases that exceed the MSV, being the P90s, P50s and P10s volumes depicted in Fig. 4.12. The suffix 's' is a reminder that these volumes are from the success cases only; the P90s implies that once the MCFS has been met, there is a 90% probability of exceeding the P90s volume.

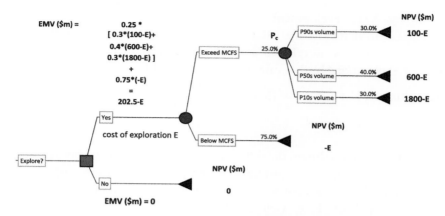

Figure 4.17 EMV based on three outcomes that exceed the MCFS.

Note that the P50s volume is slightly lower than the MSV since the distribution is positive skewed and hence the NPV of the P50s case is slightly lower than the NPV of the MSV case. This approach is very defensible since it requires that the economic evaluation is made for multiple deterministic cases that exceed the MCFS and thus considers a range of volumetric outcomes.

Case 3 approach is to work only with the P_g estimate, take the P90, P50 and P10 volumes from the distribution and estimate the NPV of developing these volumes. These will be lower volumes than the P90s, P50s and P10s which are taken from the truncated volumetric distribution, but P_g will be higher than P_c (see Fig. 4.11), so for this example assume that if P_c is 0.25 then P_g is 0.3.

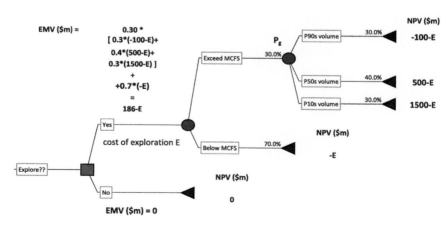

Figure 4.18 EMV based on three volumetric outcomes (no MCFS cut-off).

In Fig. 4.18, development of the P90 volume yields a negative NPV as it happens to be below the MCFS. One could choose to assume that this case would not be developed and the value would be $-E$, simply the cost of exploration. In that case, the EMV would become

$$0.3 \times [0.3 \times (-E) + 0.4 \times (500 - E) + 0.3 \times (1500 - E)] + 0.7 \times (-E)$$

$$= \$195 - E \text{ m}$$

In the examples shown the difference in the EMV estimate between the cases is $16.5 m, or approximately 8%. Of the three approaches demonstrated, the simplest is a single economic evaluation of the MSV combined with the P_c to yield a risked value on which to base a simple decision such as ranking opportunity against each other. Running development economics on multiple cases that exceed the MCFS is more rigorous, as a range of outcomes is considered and the MCFS becomes a useful target for the exploration team to consider. Using the P_g and a range of outcomes is also valid but runs the danger of including a negative NPV in the EMV estimate, by accepting the development of a small discovery which turns out to make a monetary loss.

The application of the MCFS and evaluation of the range of volumetric outcomes above the MCFS is recommended, but the key within an organisation is to apply a chosen practice to all exploration evaluations to ensure consistency.

4.2.7 Summarising the opportunity

Table 4.4 is a suggested summary of the exploration evaluation, underpinning the basis of a bid. This assumes an oil prospect but could be expanded for a gas prospect or combined oil and gas.

Table 4.4 Summary of the exploration opportunity evaluation.

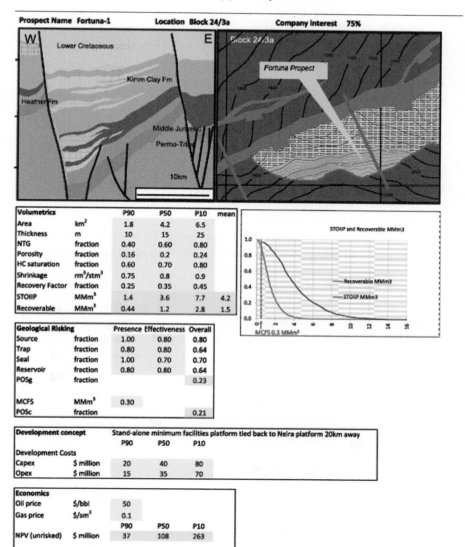

Prospect Name	Fortuna-1	Location	Block 24/3a		Company interest	75%

Volumetrics		P90	P50	P10	mean
Area	km²	1.8	4.2	6.5	
Thickness	m	10	15	25	
NTG	fraction	0.40	0.60	0.80	
Porosity	fraction	0.16	0.2	0.24	
HC saturation	fraction	0.60	0.70	0.80	
Shrinkage	rm³/stm³	0.75	0.8	0.9	
Recovery Factor	fraction	0.25	0.35	0.45	
STOIIP	MMm³	1.4	3.6	7.7	4.2
Recoverable	MMm³	0.44	1.2	2.8	1.5

Geological Risking		Presence	Effectiveness	Overall
Source	fraction	1.00	0.80	0.80
Trap	fraction	0.80	0.80	0.64
Seal	fraction	1.00	0.70	0.70
Reservoir	fraction	0.80	0.80	0.64
POSg	fraction			0.23
MCFS	MMm³	0.30		
POSc	fraction			0.21

Development concept		Stand-alone minimum facilities platform tied back to Neira platform 20km away		
		P90	P50	P10
Development Costs				
Capex	$ million	20	40	80
Opex	$ million	15	35	70

Economics					
Oil price	$/bbl	50			
Gas price	$/sm³	0.1			
		P90	P50	P10	
NPV (unrisked)	$ million	37	108	263	
EMV	$ million				31
Company share	$ million				23

This summary would be supported by more detail of the risk evaluation for the play and prospect specific risks, and may be expanded to include a range of oil and gas prices.

In this example, the volumetrics have been generated using a combination of triangular distributions (area, thickness, recovery factor, all defined using the P90, P50, P10 estimates), normal distributions (NTG, porosity and saturation, defined with mean and standard deviations) and a uniform distribution for shrinkage (defined using a minimum and maximum).

The STOIIP and Recoverable distributions shown are un-risked, so the graph shows the probability of finding the minimum quantity of hydrocarbons as 1.0 and the probability of recovering the MCFS of 0.3 MMm³ is 0.9. The P_g is estimated to be 0.23 and hence the P_c is 0.21 (=0.23 × 0.9).

The decision tree that would support this calculation is shown in Fig. 4.19, created using PrecisionTree®. The P_g is used in this case since the P90 volume is greater than the MCFS.

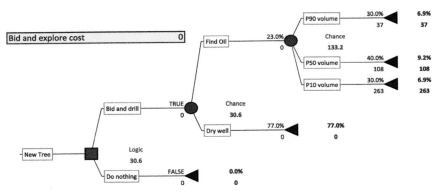

Figure 4.19 Decision Tree supporting EMV calculation.

The EMV calculated ($30.6 m) assumes no cost of bidding or exploring — it is effectively the expected monetary value of the undrilled prospect. This will form the basis of the next decision — how to formulate the bid.

4.2.8 Formulating the bid — theory and practice

The chapter so far has presented techniques to estimate the EMV of an exploration opportunity using a structured approach to estimating uncertainty in volumes and then combining the exploration risk. The techniques assume that we are able to adequately assess a realistic range of volumetric uncertainty, a reasonable probability of success of finding hydrocarbons, appropriate development plans, development and operating costs and cost of capital. It then assumes we apply the discounted cash flow method to estimate NPVs for various outcomes and use decision tree analysis appropriately to yield a prospect EMV.

These techniques allow us to be consistent in our approach, but there are many assumptions made along the way. If, and only if, all the assumptions were correct and the techniques applied rigorously, would the EMV calculation be a true reflection of the value of the opportunity. The sums are important, but the final decision on formulating the bid will be in the hands of the management team, whose insight will be added to the calculation.

The exploration bid most commonly comprises a technical commitment, a signature bonus plus other benefits, and possibly a commitment to fiscal terms (covered in more detail in Chapter 6). The technical commitment will be to carry out an amount of surveying (say 5000 km^2 of 3-D seismic acquisition), drill a number of wells (say 5 wells to a minimum depth of 3000 m). The signature bonus offered will be payable upon award of the licence. This is usually not a prescribed sum in the tender, though a minimum may be advised, and can vary hugely. For high-risk small licence areas this may be in the region of hundreds of thousands of dollars, but can run into billions of dollars, as reportedly paid by Sonangol-Sinopec in Angola ([12], IHS Markit). Other elements which may be offered in the bid include social welfare benefits, such as commitments to education or technical support for research centres. In countries that use Production Sharing Contract arrangements, signature bonuses and additional benefits are commonly offered — this is not true of the more traditional oil- and gas-producing regions.

The cost of the bid will be the sum of the elements above. Logic would dictate that the bid cost should not exceed the calculated EMV of the undrilled prospect (i.e. not deducting any exploration or signature bonus costs, $202 m in Fig. 4.17). The EMV is the probability-weighted value of all the possible outcomes evaluated. If the single exploration opportunity were exploited hundreds of times, this would be the average outcome — that is essentially the meaning of EMV; the dice is rolled many times. Therefore if the bid were equal to the EMV of the undrilled prospect the remaining EMV would be zero, and it is illogical to pursue exploration business with the expectation of zero net value.

This leads to the question of what fraction of the EMV to bid and still leave some net value in the opportunity. In the example in Fig. 4.17, a bid of say $50 m would leave an EMV of $152 m, which would be very satisfactory. However, in a competitive situation, this may not win the bid, leaving the company without the opportunity to explore. The EMV calculation is a useful guide, but deciding what fraction of this to bid becomes a matter of taste (to use a common economist's term) decided by the insight and discretion of the management team making the final decision.

This raises an important, if philosophical, point about economics and risk analysis. The sums may be calculated in a structured and consistent way, but the overprint of management's insight is more subjective, and will often adjust or even override the calculation. The technician must do the calculation, and be guided by the results, but recognise that the final decision-maker is not totally beholden to the sums.

If the level of competition is judged low, then a low fraction of the calculated EMV may win the licence award. In a highly competitive situation, it is tempting to bid close to the EMV, leaving a small margin of profit, assuming the probability-weighted outcome, while hoping for the upside in the actual outcome (find more than the P50 volume case). Note again that the EMV combines all the outcomes perceived with their associated probabilities and is the average result — hoping for the upside outcomes is just that, a hope.

The historical outcome of bidding may also influence what fraction of the EMV to bid. If bidding 20% of the calculated EMV never wins the award, then this feedback will encourage a higher fraction to be bid.

The risk tolerance of the investor will also influence the final bid. This can be represented by a utility function, which indicates the true impact of a monetary loss to the investor. Large companies are likely to have a greater tolerance to loss than a small independent investor, and may therefore be willing to place a higher licence bid. The scale of some opportunities may be outside the risk threshold of some potential bids. An understanding of the appetite for risk of one's own company and that of the competition will be helpful. The concept of utility is introduced in Chapter 10.

Other factors that will influence the bid are listed in Table 4.5, which demonstrates that the consideration for the bid is a multi-disciplinary effort.

Table 4.5 Factors influencing exploration bid value.

Factor	Description
Strategic fit	Opportunity to expand E&P business
	Opportunity to link to downstream businesses
Country risk	Political stability
	Fiscal stability
	Supply chain maturity
	Reputational threats
Technical complexity	Location (water depth, terrain and weather conditions)
	Access to existing infrastructure
Health, safety and environment	Welfare of local population
	Safety of company employees and contractors
	Environmental impact of development

Evaluating these elements is more subjective than the application of the techniques to estimate the 'technical' EMV of an exploration opportunity.

The degree of desire of management to enter into a new region or a geological play will also influence the bid, and in many instances has driven companies to bid above the EMV calculated by the methods presented in this chapter. After a discovery, this may turn out to have been a good decision, but can also lead to severe strain and possible failure of companies where the prospect does not achieve close to the EMV calculated. It has long been recognised [1] that bidders can win a licence but then suffer 'winner's regret'.

Particularly under PSC fiscal terms, the government offering the exploration licence will expect a signature bonus to be part of the bid, along with the technical elements (wells, surveys). Once the company has decided what fraction of the EMV to bid, there remains a decision on the blend of these elements. There may also be decisions to make on the profit oil split (discussed in Chapter 6).

In some cases, the government will suggest a minimum signature bonus. Otherwise, the bidder must judge how the government will evaluate the offer — does the signature bonus outweigh the technical element? In this regard, it is useful to look at the results of previous bidding round, and in some countries all bids are published to provide transparency. It may be useful to have held discussions with the relevant ministries to gain insight into the government's view. This is often part of the Gaining

Entry phase of the field life cycle, during which companies may take up a position in-country on a technical advisory basis to the government ministry or the national oil company. During this period, relationships and trust can be built, placing the bidder in an advantageous position during bid evaluation.

References

[1] E.C. Capen, R.V. Clapp, W.M. Campbell, Competitive bidding in high-risk situations, J. Petrol. Technol. 23 (1971) 641−653, 10, 2118/2993-PA.
[2] R.E. Megill, An Introduction to Risk Analysis, in: O.K. Tulsa (Ed.), second ed., PennWell Books, 1984.
[3] P.A. Allen, J.R. Allen, Basin Analysis. Principles and Applications, Blackwell Scientific, 1990.
[4] P. Ringrose, M. Bentley, Reservoir Model Design − A Practitioner's Guide, Springer, 2015.
[5] SPE, et al., Petroleum Resource Management System, 2018, ISBN 978-1-61399-660-7.
[6] P.L. Bernstein, Against the Gods: The Remarkable Story of Risk, Wiley, 1998.
[7] Crystal Ball Software. https://www.oracle.com/.
[8] @RISK Software. https://www.palisade.com/.
[9] MCApp Software. https://jumpingrivers.shinyapps.io/mcapp/.
[10] A. Hurst, G.C. Brown, R.I. Swanson, Swanson's 30-40-30 rule, Am. Assoc. Pet. Geol.Bull. (AAPG) 84 (No. 12) (December 2000).
[11] V. Singh, E. Izaguirre, I. Yemez, H. Stigliano, Establishing Minimum Economic Field Size and Analysing its Role in Exploration Project Risk Assessment: Three Examples, Search and Discovery Article 40827, 2016.
[12] IHS Markit, Global Insight Perspective, May 12, 2006.
[13] https://palisade.com.

CHAPTER 5

Appraisal planning

Contents

5.1 Reservoir appraisal and project assessment in context	101
5.2 Justifying reservoir appraisal assuming perfect information	103
5.2.1 Appraising for comfort	109
5.2.2 Appraising to prove commerciality	109
5.2.3 Appraising to add hydrocarbon volume	110
5.2.4 Appraising to avoid sub-optimal expenditure	111
5.3 Incorporating imperfect information	114
5.3.1 Bayesian revision of probabilities	114
5.3.2 Imperfect information from a seismic survey	118
5.4 Reservoir appraisal planning	122
5.4.1 Target of appraisal — tornado diagrams	122
5.4.2 Appraisal tools	123
5.4.3 Impact of appraisal on project schedule	126
5.4.4 A medical example of Bayesian revision	128
References	131

5.1 Reservoir appraisal and project assessment in context

Chapter 1 introduced the appraisal phase of the project lifecycle as the step that delineates reservoir volumes, once an exploration discovery has been made. This chapter will address appraisal in the same context.

The objective of appraising for improved understanding of reservoir volumes is to enable value to be added to the project by making an informed and thus better decision on the specifics of the field development. The decisions concern right-sizing facilities, specifying sales contract terms, appropriate scheduling, or sometimes stopping the project from progressing further.

However, the term 'appraisal' may appear in another context — a step in the Stage Gate process. This is not specific to reservoir evaluation, but more generally used in project control and planning.

NASA practised the concept of phased development in the 1960s with phased project planning or phased review processes, intended to break up the development of any project into a series of steps that could be individually reviewed in sequence. The Stage Gate process was formally introduced into project planning practices by Prof. Robert G. Cooper, who applied the technique to conceiving, developing and launching new products for industries including chemical manufacturing and the oil industry. He patented and trademarked the Stage-Gate Idea-to-Launch Process [1].

Fig. 5.1 illustrates a typical Stage Gate process in which the second step is named 'Assess' — alternatively sometimes called 'Appraise'.

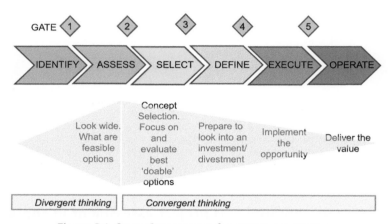

Figure 5.1 Stage Gate process for project progression.

The Stage Gate process is practised to check project progress at various points. Tests are performed at the Gates, to determine whether the work performed is mature enough to allow the project to pass to the next stage. If not, the project may be stopped, or more work performed at that stage, or perhaps recycled to a previous stage. The process is used as a control to prevent projects from progressing with insufficient analysis, reducing the chance of subsequent disappointment or failure. The Gatekeeper role is often performed by a group external to the project team, so that an objective and dispassionate view can be taken, and a set of standard tests can be applied.

The stages are summarised in Table 5.1. Note that at the various stages the composition of the team often changes, so effective communication in the handover from one stage to the next is critical to success.

Table 5.1 Description of Stage Gate steps.

Stage	Description
Identify	Identify an investment opportunity.
Assess (or Appraise)	Consider various options to commercialise the opportunity. Keep the options open, think wide and do not reject possibilities too soon. Demonstrate that at least one option is technically and commercially viable, i.e. it is possible to do, conforms to HSSE (Health, Safety, Security, Environment) standards, satisfies regulatory requirements and passes basic economic hurdles.
Select	From the options generated, start to filter these by ranking them on technical, HSSE and commercial merit until at the end of this stage, one concept for development can be selected.
Define	Work up the detail of the selected option by performing detailed engineering design and cost estimating so that at the end of this stage a decision can be made to approve and finance the project by issuing the Final Investment Decision (FID) to sanction the project progression.
Execute	Build the project. This often involves construction by third parties, so the oil company outsources a large part of the work to contract groups. The project team composition becomes quite different.
Operate	The built project is commissioned and subsequently operated. Commissioning is often a joint responsibility of the oil company and the construction contractor. At the end of this stage, the project is handed back to oil company operators, so the team composition changes again.

5.2 Justifying reservoir appraisal assuming perfect information

Returning to Appraisal in the context of improved understanding of the reservoir, there is a similarity to the Stage Gate process in the handover from one team to another, in relation to the roles of the exploration team and the subsurface field development team. Various oil companies use different approaches with regard to who is responsible for reservoir appraisal planning. Post-discovery, some choose to pass the reservoir appraisal activities to the development team while others prefer the exploration team to continue with reservoir appraisal.

The natural preference of an exploration team is to prove the presence of hydrocarbons, and its skill set supports this. If the team's success is judged by making discoveries, then the tendency to strive to prove up more volume will be reinforced. If the budget for the appraisal activity rests with the exploration team, then it becomes likely that budget will be allocated to finding more hydrocarbon volumes.

If it is accepted that the objective of appraisal is to refine the understanding of the reservoir character in order to add value by selecting an appropriate development plan, then it becomes reasonable to pass the responsibility for appraisal to the development team.

Post-discovery, the most common significant elements of the reservoir character that require improved understanding are reservoir size, complexity and well productivity. To add value to the project it is not necessary that appraisal must add further volumes to those attributed at discovery, as the following example will demonstrate.

Suppose a single exploration well (X-1) has made an offshore discovery. At the well location, the seismic response for top reservoir, measured in time, can be directly linked to depth from the logging and drilling data, so that the sonic velocity of the overburden is calculated at this location. However, away from this location, the overburden velocity may vary, giving rise to uncertainty in the extent of the reservoir, as depicted in Fig. 5.2 where three contours have been mapped to represent the P90, P50 and P10 estimates.

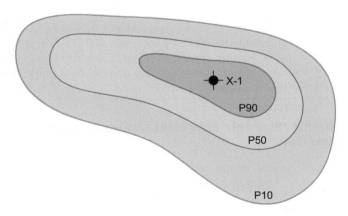

Figure 5.2 Uncertainty in reservoir extent post-discovery well X-1.

If a development were to progress with no further information, then a decision would be required to size the facilities (platform, processing plant, export pipeline). The typical decision would be to design for the P50 case, since aiming at the mid-point of the distribution feels most comfortable. Noting Swanson's rule introduced in Chapter 4, the P50 represents the average of approximately 40% of the outcomes.

Constructing facilities for the P10 case carries the risk of overspending on the development — Swanson's rule suggests that the P10 case is representative of 30% of the outcomes, so there is a 70% chance of overspend. On the other hand, it may provide the option to tie in future near-field discoveries, and thus carry so-called option value.

Constructing for the P90 case carries the risk of under-sizing the facility and being obliged to defer production due to limited throughput capacity, and thus reduce the Present Value (PV) of the revenue stream. If the P90 case represents 30% of the outcomes, there is a 70% chance of losing value due to deferral.

So, assuming that the facility is designed and costed for the P50 field size, the net present value (NPV) of the project can be estimated for the actual field size, found to be either the P90, P50 or P10 case. If the project's NPVs are estimated as $20, $200, $400 m these outcomes can be weighted according to Swanson's rule at 30%, 40%, 30% to calculate the EMV of the project:

$$\text{EMV} = (30\% \times \$20\text{ m}) + (40\% \times \$200\text{ m}) + (30\% \times \$400\text{ m}) = \$206\text{ m}$$

While this is a respectable EMV, in the case of a P90 field size, the NPV of the project is only marginally positive, and if the field size were slightly smaller, the project would most likely return a negative NPV. The following analysis will investigate whether the project value could be improved by appraisal of the field size before committing to the facilities design.

Let us assume that reservoir appraisal involves shooting a 3-D seismic survey aimed at improving the uncertainty in mapping of the field, and that it can clearly distinguish which of the above three oil-water contact contours is the ground truth. This implies that the seismic provides what is termed **perfect information**, meaning that if the seismic indicates, for example, the P90 case, then the field is indeed that size. The faithfulness of seismic information depends upon the quality of the acquisition, processing and interpretation — a combination of tool performance and human ability to interpret data correctly.

If the perfect information provided the knowledge of the field size, the development plan could then be appropriately sized. The value of the P90 case above would be improved by building a smaller facility and avoiding over-expenditure. The value of the P50 case would remain the same as the calculations in Fig. 5.2 which assumed constructing mid-sized facilities. The value of the P10 development would increase by installing a larger facility and accelerating production compared to being constrained by the capacity of facilities designed for the P50 case. It is assumed that the additional cost of the larger facility is more than offset by the increased PV of the accelerated oil revenue.

Fig. 5.3 shows a decision tree that allows the calculation of the maximum value of the 3-D seismic, which is assumed to provide perfect information. This figure shows how the tree could be drawn by hand, using the cost of appraisal as being $-A.

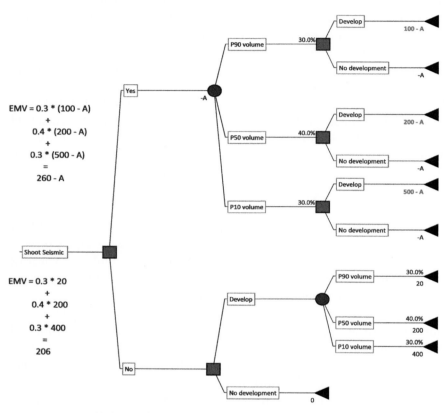

Figure 5.3 Hand-drawn decision tree to estimate value of appraisal information.

P90, P50, P10 weighted probabilities are assigned using Swanson's rule, and where a decision is required the higher value has been coloured green. Note that 100-A always exceeds -A, regardless of A.

The EMV of the project with the seismic appraisal is $(260-A) m, while the value of the project with no appraisal and immediate development is $206 m. The maximum value of the appraisal activity is thus $(260−206) m = $54 m. It is a general statement that

Value of information (VOI) = EMV of project with information − EMV of project without information

Fig. 5.4 shows the PrecisionTree version of this example. It is convenient to insert a cost of appraisal into the spreadsheet as highlighted in yellow, and refer to this cell below the branch following the decision to appraise. In this example the cost of appraisal is set to $0 m but if set to say $-20 m then all the net payoffs following that decision will be reduced by that amount and the EMV of the Yes (shoot seismic) branch would reduce to $240 m.

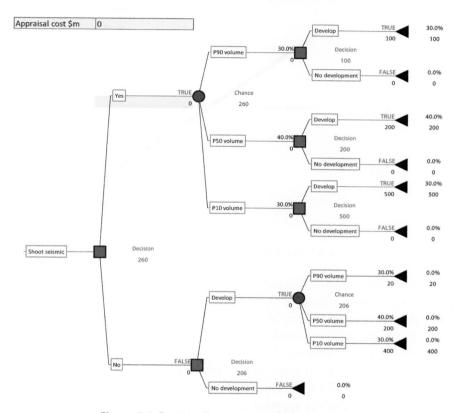

Figure 5.4 PrecisionTree version of appraisal VOI.

The software has allowed the problem to be organised consistently and presented clearly, but the same calculation can readily be done by hand, and drawing out the logic of the tree by hand before applying the software it is strongly encouraged.

In the above example, it is simple to see that the maximum VOI is the difference between the EMV of the project with the information less the EMV of the project without the information, a difference of $54 m.

The software can perform a sensitivity analysis on any of the assumptions in the tree, and Fig. 5.5 shows the result of varying the cost of the seismic.

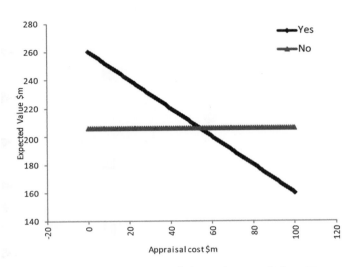

Figure 5.5 Sensitivity analysis on the appraisal cost.

As the cost of shooting the seismic increases, the EMV of the project reduces, up to the breakeven of $54 m cost. If the seismic cost exceeds this then the appraisal starts to destroy value. The maximum VOI in this case is $54 m. It is important that this is not considered as an appraisal budget. In the conventional sense, having been granted a budget it is common to then spend it all, in case future budgets are cut back. This is not the meaning of a VOI sum of money; it is not a budget.

The objective of the appraisal is to reduce the identified uncertainty (in this case areal extent of the discovery) as efficiently as possible. If seismic can reliably resolve the uncertainty for $10 m then it is predicted to add $44 m of value to the project. If it costs $54 m then it is not adding any value, and should not be justified.

Note that the VOI is based on the NPVs of the development cases, net of royalty and taxes. For consistency, the appraisal VOI should also be corrected for any fiscal relief available. The fiscal allowances for appraisal costs may depend on the outcome of the project. If no development ensues then appraisal costs may not be offset against the revenues from the development project. However, if a development does take place, then an opportunity may arise to offset the appraisal costs against the development project revenues. The local fiscal terms will dictate the treatment of appraisal costs, but such costs will generally be allowable against corporation tax.

Clearly, care is required to calculate the true net of tax cost of appraisal from a fiscal perspective. In addition, the cost implications of delay to the project while appraising need to be considered, and this will be addressed in Section 5.4.3.

5.2.1 Appraising for comfort

In the above example, the intention of shooting the seismic is to reduce the range of uncertainty in order to make a better decision on sizing the facility for field development. Unless data collected during appraisal is going to assist a future decision then it is not justifiable to spend money gathering it. It becomes an indulgence to spend appraisal money to satisfy curiosity or to provide a degree of comfort — the appraisal activity should influence a future decision.

5.2.2 Appraising to prove commerciality

The minimum hydrocarbon volume proven up by an initial discovery well may not exceed the MCFS (minimum commercial field size — recall this from Chapter 4). Further appraisal may be undertaken to prove whether the estimated volume exceeds the MCFS, but this does not optimise the development plan — it only demonstrates that the discovery passes the hurdle of making an economic breakeven when developing with facilities designed for the MCFS. The value of such appraisal is not the same as value gained by enabling optimisation of the development plan, as discussed so far.

However, appraisal that proves the MCFS may add significant value to a discovery if it then becomes an attractive sales proposition and the owner can sell the asset, or if proving commerciality allows the owner to raise funds for development. It may also allow the hydrocarbon volumes to progress upwards in the classification of resources. For example, using PRMS definitions [2] it may allow the classification to progress from the Contingent Resources class 'Development Not Viable' to 'Development Pending', and thereby improve analysts', shareholders' or purchaser's perception of value.

If the stated objective of appraisal is to prove the MCFS, then this should drive the form of appraisal as illustrated in Fig. 5.6. Here the appraisal well is located to prove an MCFS by targeting a location just updip of a hydrocarbon water contact that would correspond to the MCFS. The example assumes a simple structure and a single contact. If the well finds water then the discovery may be considered as a commercial failure, but at least that has been established at the cost of a single well. The example suggests that MCFS depends on areal extent of the field, but the hurdle to

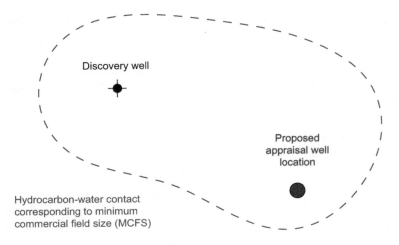

Figure 5.6 Appraisal to prove MCFS.

commerciality may be due to other factors such as well productivity, reservoir permeability or sealing nature of mapped faults. The form of appraisal should be designed to address the key uncertainty, and rather than being another well could be a production test on the discovery well, an interference test between wells or 3-D seismic acquisition.

5.2.3 Appraising to add hydrocarbon volume

Unless trying to prove an MCFS as in Section 5.2.2, it is unlikely that appraisal activity can be justified on the basis of proving additional hydrocarbons. Some decision-makers consider that an appraisal well which finds additional hydrocarbon represents a success, and consider a dry well (water-bearing) to be a failure. This is an incorrect perception of appraisal. Fig. 5.7 illustrates an exploration well followed by three appraisal wells, the first two of which prove additional hydrocarbons, but the third finds water and demonstrates that the field is below the MCFS. This is a very inefficient form of appraisal, since drilling well A-3 first would have

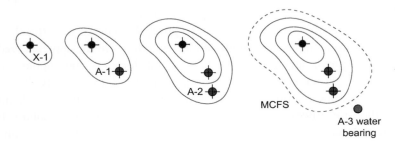

Figure 5.7 Step-out appraisal.

demonstrated the field being less than the MCFS, for the cost of one appraisal well only. The dry well A-3 is the most successful appraisal well, as it informs the decision not to progress further with this discovery. After drilling well X-1, such a discovery is sometimes termed a 'teaser' since it has encouraged the drilling of a series of appraisal wells, the first two adding volume and raising expectations, but the last of which is the let-down. Being drawn into this approach to appraisal is not effective. The objective of appraisal should be to reduce uncertainty to allow a better subsequent decision to be made, in this case no further activity. Arguably, the owner should have tried to sell the asset immediately after discovery.

5.2.4 Appraising to avoid sub-optimal expenditure

At the start of Section 5.2 it was stated that a benefit of appraisal is to avoid both over-expenditure by oversizing the facilities (simply spending more than necessary) and under-expenditure by undersizing facilities (and incurring production deferral). The over- or under-expenditure may be termed sub-optimal expenditure (SOE). In the ideal case, if the exact size of the field were known, appropriate-sized facilities would be constructed, and no SOE would be incurred.

Using this framework, this section will demonstrate that successful appraisal reduces SOE, which can be achieved by finding either smaller of larger volumes than the P50 estimate post-exploration, as long as the uncertainty range is reduced. The approach can be quantitative, but will remain qualitative in this section.

If the exact size of the field were known, the appropriate development (wells, facilities) could be planned and costed. Fig. 5.8 (a) illustrates the relationship between development cost and field size, in broad terms, while (b) recognises that there will be certain break points at which the

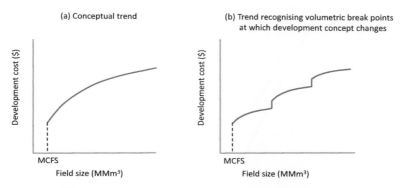

Figure 5.8 Relationships between development cost and field size: (a) Conceptual trend and (b) more realistic trend with concept design break points.

facilities concept may change with field size, from say a satellite platform to a stand-alone processing platform to a fully integrated platform with a dedicated export system.

Fig. 5.9 illustrates the range of volumetric uncertainty pre-appraisal and assumes that wells and facilities are designed for the P50 case. If the volume realised is close to the P50 then there will be no under- or over-expenditure, and some flexibility is assumed in minimising SOE if the volume is close to the P50.

If the volume realised is significantly greater than the P50 then the under-sized facility will result in production deferral and cause a loss of value compared to building the right-sized facility. This is the SOE due to under-sizing, which increases as the volume realised increases. On the other hand, if the volume realised is significantly less than the P50 then the SOE due to over-sizing is the loss of value caused by spending too much on the facility compared with the optimum sizing.

Calculation of the SOE due to over-expenditure and under-expenditure requires the interaction of the subsurface team (volumes, production profiles), facilities engineers (development costs) and the economists (NPV for each combination), which is not an insignificant effort. But if the SOE values are calculated as a function of volume and we have a probability distribution for the volumes, then the risked SOE is the product of the SOE at each volume outcome multiplied by the probability of occurrence. The risked SOE is represented in Fig. 5.9 as the sum of the shaded areas. This has the unit of currency ($) and can therefore be considered quantitative.

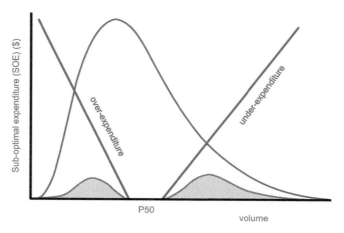

Figure 5.9 Sub-optimal expenditure (SOE) in designing for the P50 pre-appraisal volume.

Fig. 5.10 illustrates that when successful appraisal reduces the range of volume uncertainty, the risked SOE (the sum of the grey areas in each plot) is reduced in all cases, whether the post-appraisal P50 volume stays similar, moves higher or moves lower. In both the high and low cases, a more appropriate facility size can be constructed.

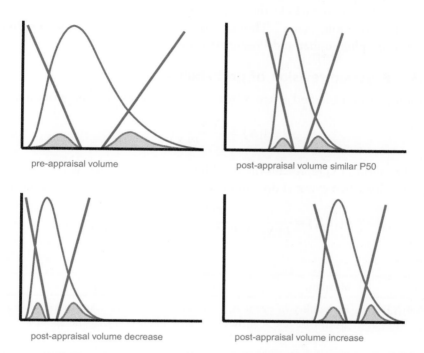

Figure 5.10 Impact of appraisal on reducing sub-optimal expenditure (Axes are as in Fig. 5.9).

This approach reinforces the point that the value of reservoir appraisal is derived from narrowing the range of volumetric uncertainty, allowing a better decision on facilities sizing. The majority of the additional value usually comes from identifying whether the ground truth lies on the high or low side of the pre-appraisal range.

Using this technique in a quantified manner requires significant effort, and the decision tree approach using a limited number of outcomes is often preferred as being quicker and more pragmatic.

5.3 Incorporating imperfect information

The chapter has so far assumed that the information provided through appraisal is wholly reliable, referred to as 'perfect information'. However, this ideal is not necessarily true as tests can be unreliable and this may significantly reduce the value of information (VOI). The VOI calculation can be adapted to include the impact of imperfect information using a Bayesian revision, after Thomas Bayes, an 18th-century English statistician, philosopher and Presbyterian minister.

5.3.1 Bayesian revision of probabilities

Noting an event A and a test X, then Bayes' theorem can be written as

$$P(A|X) = \frac{P(X|A) \cdot P(A)}{P(X)}$$

The term $P(X)$ represents the probability of a positive test when event A occurs plus when event A does not occur, and may be expanded as follows:

$$P(A|X) = \frac{P(X|A) \cdot P(A)}{P(X|A) \cdot P(A) + P(X|notA) \cdot P(notA)}$$

where

Term	Meaning	Simplified meaning
$P(A\|X)$	Probability of event A, given that test X shows event A	This is what we want to know — how likely is it that the event is real with a positive test result
$P(X\|A)$	Probability of a positive test, given event A occurs	Probability of a true positive
$P(A)$	Probability of event A	Known as the a priori *probability* — the ground truth probability
$P(X)$	Probability that test X shows event A	This is the sum of all the positive results — the true positives plus the false positives
$P(X\|notA)$	Probability of a positive test if event A does not occur	A false positive result
$P(notA)$	Probability of event A does not occur	1-P(A)

In the above meanings, the term ground truth is used to refer to information provided by direct observation, empirical evidence, as opposed to an estimate or information provided by inference. In simple terms it is the reality. In some fields, such as medicine, this is termed the prevalence.

An example is useful here, but please be aware that the data used is assumed for this purpose and should not be taken as necessarily accurate. The methodology is, however, robust.

Insurance company statistics show that 1 in 20 houses suffers from a break-in by an intruder within a period of a year. For a house fitted with an alarm, there is a 90% probability that the alarm sounds when an intruder enters. However, there is a 20% chance that in a year the alarm trips by mistake and sounds, even though there is no intruder (e.g. pets, gales, power trips). So if you hear a house alarm, what is the probability that there is actually an intruder? Bayes' theorem can help to answer this question.

The a priori *probability* of a house break-in by an intruder in a year $P(A) = 0.05$. The a priori probability of not having a break-in $P(notA)$ must therefore be 0.95.

House alarms are understood to be partly unreliable in that they sound the alarm in 90% of cases when there is an intruder (so $P(X|A) = 0.90$), but they also trip and sound an alarm 20% of the time, even though there is no intruder (so $P(X|notA) = 0.20$).

Bayes' theorem can be applied to estimate the question we would like to answer — what is the probability of there actually being an intruder when the alarm sounds, $P(A|X)$.

$$P(A|X) = \frac{P(X|A) \cdot P(A)}{P(X|A) \cdot P(A) + P(X|notA) \cdot P(notA)}$$
$$= \frac{0.9*0.05}{0.9*0.05 + 0.2*0.95} = \frac{0.045}{0.045 + 0.19} = \frac{0.045}{0.235}$$
$$= 0.191 = 19.1\%$$

The 0.191% is termed the *posterior probability*. The probability of the alarm sounding is the sum of all the positives, the bottom line of the equation (23.5%), but once the new knowledge becomes available that the test is not fully reliable, the posterior probability of actually having an intruder, given the alarm sounds is revised to 19.1%.

The equation can be simplified down to

$$(A|X) = \frac{P(\text{true positive})}{P(\text{true positive}) + P(\text{false positive})}$$

Or even further to

$$(A|X) = \frac{P(\text{true positive})}{P(\text{any positive})}$$

If the formulae do not convey the message, then it can be helpful to construct a form of Venn diagram to calculate the posterior probabilities. The coloured area in Fig. 5.11 contains all possible outcomes (the total event space). Prior probabilities are plotted on the x-axis (not to scale) and the reliability on the y-axis.

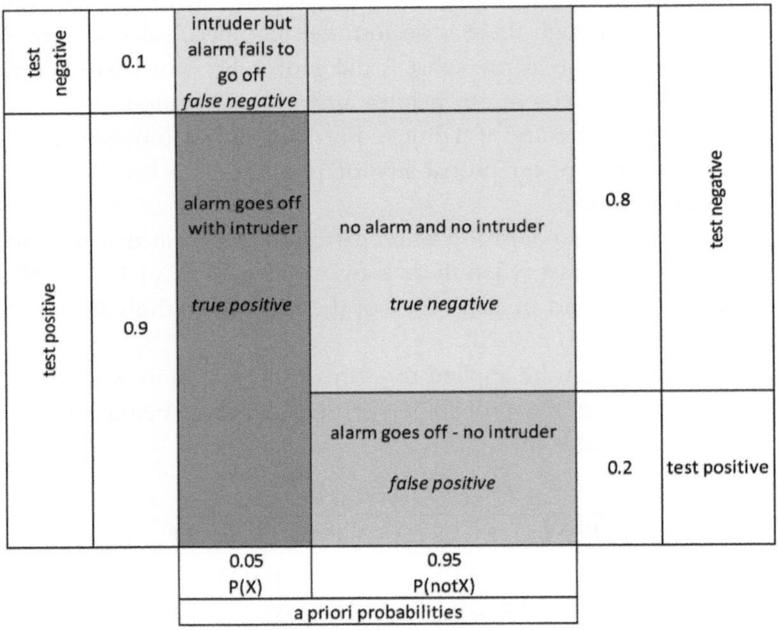

test negative	0.1	intruder but alarm fails to go off *false negative*			
test positive	0.9	alarm goes off with intruder *true positive*	no alarm and no intruder *true negative*	0.8	test negative
			alarm goes off - no intruder *false positive*	0.2	test positive
		0.05 P(X)	0.95 P(notX)		
		a priori probabilities			

Figure 5.11 Bayes' theorem for house break-in as a Venn diagram.

The figure also identifies the four outcomes, true positive, false positive, true negative and false negative, making the presentation of the problem more generic.

$$(A|X) = \frac{P(\text{true positive})}{P(\text{any positive})} = \frac{\text{dark pink area}}{\text{dark pink area} + \text{light pink area}}$$

The result may be somewhat surprising (or even alarming!), and may discourage you from reporting a sounding alarm to the authorities. Such action would skew statistics of course since the incident would not be recorded and therefore not become part of the statistical database. This is always an issue with a data set (the unknown unknowns), but it is in the public and insurance company interest to report a house alarm.

A further way to approach the application of Bayes' theorem is to use a probability tree, as shown in Fig. 5.12.

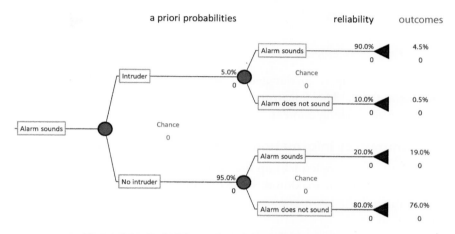

Figure 5.12 Probability tree approach to Bayes' theorem.

In this case the probability the alarm sounds is (4.5% + 19.0%) = 23.5% but the alarm sounding is only a true positive 4.5% of the time, so that the probability of there being an intruder given the alarm sounds is only 4.5%/23.5% = 19.1%.

In general, when the a priori probability of a positive result is low and the probability of a false positive is high, the positive test result can be very misleading. This is very significant in the world of medicine, and good examples for this discipline are illustrated by Kit Yates [3].

A very convenient tool in PrecisionTree® is the ability to calculate a Bayesian revision of probabilities. Having created the tree in Fig. 5.12 using the prior probabilities and the reliabilities, the Bayesian Revision option is applied and the revised probabilities are generated as shown in Fig. 5.13.

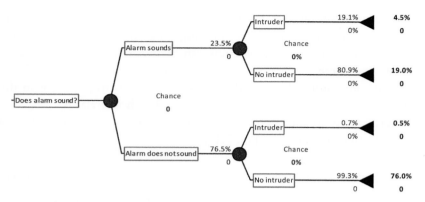

Figure 5.13 Bayesian revision of probabilities using PrecisionTree®.

Further reading is encouraged in an Economist article 'How science goes wrong' [4]. The following section applies Bayesian revision in an E&P decision-making context.

5.3.2 Imperfect information from a seismic survey

The following example picks up the decision to shoot seismic as presented in Fig. 5.4, where the maximum value of shooting seismic was calculated to be $54 m, and the assumption was that the seismic would provide perfect information.

Perfect information implies that if the seismic information indicates the reservoir to be the P90 volume, then this will be the case, with the same implications for the predicted P50 and P10 cases. This assumption is not necessarily valid, as the reliability of the seismic depends upon the quality of data acquisition, processing and subsequent analysis by the seismic interpreter, who is only human.

Let us assume that based on experience of actual outcome against prediction, the combination of these steps provides 80% reliability, so that

- when the interpretation is the P90 volume it is correct 80% of the time, but the other 20% of the time the ground truth is that it is the P50 volume
- when the interpretation is the P10 volume it is correct 80% of the time, but the other 20% of the time the ground truth is that it is the P50 volume

- when the interpretation is the P50 volume it is correct 80% of the time, but 10% of the time the ground truth is that it is the P90 volume and the other 10% of the time the ground truth is that it is the P10 volume

The a priori probabilities for the P90, P50, P10 cases are 30%, 40%, 30%, applying Swanson's rule. We now have the reliabilities of the interpretation being correct, which allows revised probabilities to be estimated using the Bayesian revision. This can be done using the mathematical approach, probability tables or a Venn diagram. The latter is the most visual and is presented in Fig. 5.14.

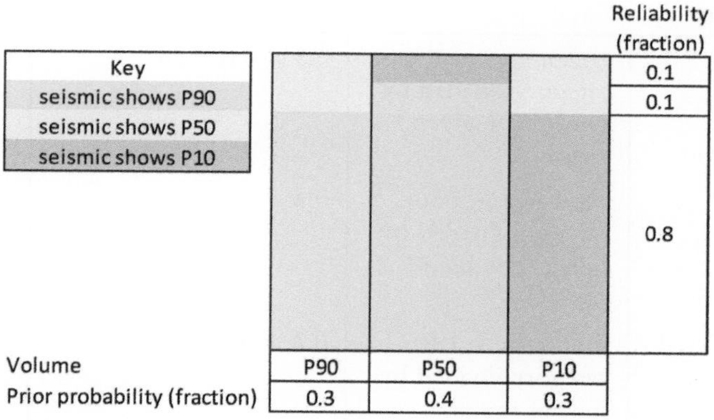

Figure 5.14 Venn diagram for Bayesian revision of probabilities.

The revised prior and posterior probabilities can now be calculated as shown in Table 5.2.

In calculating the Bayes' revised probabilities, the probability the seismic indicates the P90 is the sum of all occurrences within the Venn diagram that show the P90. However, the probability that the ground truth is the P90 when the seismic indicates P90 is the part of the Venn diagram in which the seismic indicates P90 when it is P90 divided by the sum of all occurrences within the Venn diagram that show the P90 (Table 5.2).

Table 5.2 Bayesian revision of prior and posterior probabilities.

Revised prior probability	In Venn diagram	Calculation	Result
Interpret P90	Sum of the green areas	$0.3 \times 0.8 + 0.4 \times 0.1$	0.28
Interpret P50	Sum of the blue areas	$0.4 \times 0.8 + (0.1 + 0.1) \times 0.3 + (0.1 + 0.1) \times 0.3$	0.44
Interpret P10	Sum of the pink areas	$0.3 \times 0.8 + 0.4 \times 0.1$	0.28

Posterior probabilities	In Venn diagram	Calculation	Result
Chance of actual P90 given you interpret P90	Green area in P90 category divided by sum of the green areas	$0.3 \times 0.8/0.28$	0.857
Chance of actual P50 given you interpret P50	Blue area in P50 category divided by sum of the blue areas	$0.4 \times 0.8/0.44$	0.727
Chance of actual P10 given you interpret P10	Pink area in P10 category divided by sum of the pink areas	$0.3 \times 0.8/0.28$	0.857

The decision tree can now be updated with the revised probabilities to estimate the EMV after shooting seismic when the information is only 80% reliable, as shown in Fig. 5.15. The net payoffs have been adjusted to reflect the fact that developing a P90 volume with a P50 facility would incur significant over-expenditure, and developing a P10 volume with a P50 facility would defer production and incur a loss of value compared to developing with the P10 facility.

Appraisal planning 121

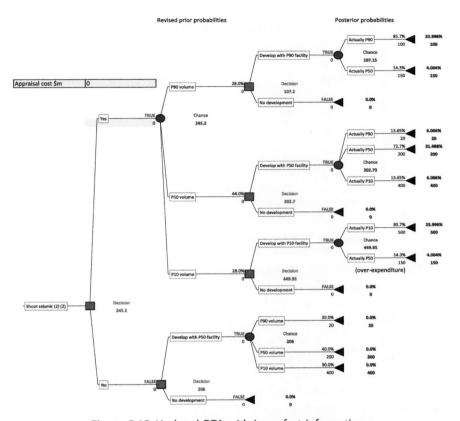

Figure 5.15 Updated DTA with imperfect information.

Given these revisions, the maximum VOI of the seismic is now $39.2 m. Recall that if the seismic information was assumed perfect the maximum VOI was $54 m. If seismic costs are $10 m as previously assumed the added value of the seismic has dropped from $44 to $29 m — a significant reduction. Those familiar with the reliability of seismic interpretation may consider 80% reliability to be fairly high. Once this evaluation process is established, it is worthwhile to run a sensitivity on the impact of the reliability on the maximum VOI, which has been run for this example and presented in Fig. 5.16. This demonstrates that the VOI can reduce rapidly as the reliability drops, and can provide a guide to whether the information should be gathered at all.

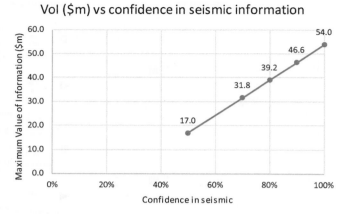

Figure 5.16 Sensitivity of VOI to reliability of information.

There may be other advantages to shooting the seismic for future reservoir management, which have not been accounted for in this analysis. For example, a baseline 3-D seismic survey prior to development may be extremely valuable to compare against later 3-D shoots and use 4-D seismic interpretation techniques to track fluid movement by investigating the difference in responses over time. If this is seen to be significant, the anticipated value may be added to the terminal nodes. It could also simply override the analysis if viewed as essential to the future project. This would be another example of performing the analysis but not becoming a slave to the result.

5.4 Reservoir appraisal planning

The chapter has so far introduced methods of justifying appraisal activity to reduce uncertainty and allow better future development decisions to be made. This section will discuss the types of reservoir appraisal activity that can be used to meet this objective.

5.4.1 Target of appraisal — tornado diagrams

It is important to be clear what information will most effectively reduce reservoir uncertainty. This depends on the root cause of the uncertainty being faced. When considering recoverable volume uncertainty, the individual elements of this are well understood, and are listed in the 'tornado diagram' presented in Fig. 5.17. This plot was created from the output of a Monte Carlo simulation predicting ultimate recovery (UR), in this case using @RISK®. As the simulation progresses, the software calculates the correlation between the change in each input variable and the change in the mean of the output, in this case UR. The ranking of these coefficients indicates which variables influence the uncertainty in the output most, as well as the direction of their effect. In this example, all high inputs lead to higher outputs, thus positive correlations. This would not always be true, for example, an increase in costs will reduce project NPV.

The tornado diagram is useful to guide the appraisal activity. The example is typical of the ranking of inputs at the appraisal stage where net rock volume (area x thickness x net:gross) commonly dominates. Appraisal should therefore aim to efficiently reduce uncertainty in these variables, starting with reducing areal uncertainty. Gathering fluid samples to reduce uncertainty in the shrinkage factor appears relatively unimportant.

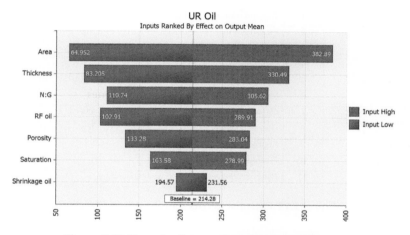

Figure 5.17 Tornado diagram for volumetric estimate.

The above ranking is not true throughout the field lifecycle. In a mature reservoir, the key variable driving uncertainty in remaining recoverable volumes may be entirely different, such as residual oil saturation. The tornado plot tool is applicable at all stages of the lifecycle to guide the target of the reservoir appraisal activity.

5.4.2 Appraisal tools

Almost any variable that influences the estimate of resource and recoverable volumes could be the target of reservoir appraisal, and Fig. 5.18 illustrates some of these. Appraisal activities should aim to reduce uncertainty in

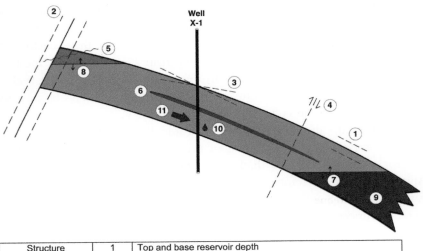

Structure	1	Top and base reservoir depth
	2	Fault position
	3	Top reservoir dip
	4	Intra-reservoir faulting
Stratigraphy	5	Erosion
	6	Net-to-gross ratio and extent of shales/barriers
	7	Oil-water contact
	8	Gas-oil contact
	9	Aquifer extent and permeability
Dynamic	10	Fluid type
	11	Reservoir deliverability and well productivity

Figure 5.18 Potential targets for reservoir appraisal.

the most significant variables as efficiently as possible, and the tornado diagram is a guide to ranking these, in addition to the experience of the subsurface team. Some simple modelling may be justified to understand the influence of dynamic properties such as reservoir permeability, well productivity and reservoir sweep efficiency during displacement of oil by water or gas.

The reduction in uncertainty in these reservoir variables can be addressed using the tools suggested in Table 5.3, noting that the time and costs are approximate and depend on the field location.

Table 5.3 Reservoir appraisal tools and interpreted information.

Tool	Subset	Potential information	Timing (months)	Cost ($m)
Seismic	Reflectors	Top and base reservoir depth Intra-reservoir horizons	24	10
	Amplitudes	Fluid contacts Reservoir properties		
Wells	Cuttings	Lithology	6	10–50
	Core	Lithology Porosity, permeability		
	Wireline logs	Lithology Porosity, permeability		
	Fluid samples	Hydrocarbon fluid properties Impurities, contaminants		
	Pressures	Fluid contacts Communication with other accumulations		
	Production test	Productivity Index Permeability Reservoir boundaries Wellbore damage		
	Injection test	As above but less environmental impact and less radius of investigation		
Analogues	Complexity index	Typical recovery factors	2	0.1
	Pressure cells	Aquifer strength		

In oil field development, a common uncertainty which influences the selection of the drive mechanism is the nature of the underlying aquifer. If an aquifer is large (some 100 times larger than the size of the hydrocarbon accumulation) and permeable, then it is likely that it will provide a natural aquifer water drive and deliver good recoveries without the need for investment in water injection wells and facilities. The size and strength of an aquifer is, however, elusive information which is only revealed after significant production from the reservoir, and is unlikely to become apparent from a short well test typically flowing for just 24 hours. Information from neighbouring producing fields, if available, is informative, but in a newly developing area this uncertainty often remains. Managing this uncertainty will be discussed further in Chapter 6.

5.4.3 Impact of appraisal on project schedule

Reservoir appraisal by gathering information using the techniques discussed in Section 5.4.2 often requires significant time to plan, execute, interpret new data and adapt decisions according to the outcome. For example, 3-D seismic acquisition, data processing, acquisition and interpretation may take up to two years from the start of planning, and possibly longer if permitting is required. While the decision using the resulting information may improve the value of the development project, the start of the development project becomes delayed.

In justifying the appraisal using a VOI approach, it is reasonable to include the effect of the delay to project development in the calculation, since the project deferral is another indirect cost of the appraisal activity.

When deciding whether to gather appraisal information, the reference date for the economic evaluation should be the time at which money is first spent on the appraisal. For consistency in calculating PVs, all costs and revenues for appraisal and any subsequent development should be corrected to the same reference date using the discounting method. It is most common that the NPV of the development project is estimated with a reference date at the beginning of expenditure on development, which will be at some future time compared with the point at which appraisal would commence.

The influence of timing can be incorporated in one of two ways for a development project that is preceded, for example, by 2 years of appraisal activity as outlined in Fig. 5.19.

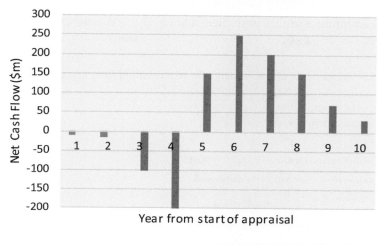

Figure 5.19 Development project preceded by appraisal activity.

Table 5.4 shows a simple way to evaluate the NPV impact of the appraisal by taking the difference in NPV between doing the appraisal followed by development [A + D] or proceeding immediately with development [D]. This approach is sometimes termed with/without economics.

Table 5.4 NPV including effect of development deferral due to appraisal.

	Year	1	2	3	4	5	6	7	8	9	10
	Appraisal net CF ($m)	-10	-15								
	Development project net CF ($m)			-100	-200	150	250	200	150	70	30
With appraisal	[A+D] net CF ($m)	-10	-15	-100	-200	150	250	200	150	70	30
	[A+D] NPV ($m)	257.8									
No appraisal	Development project net CF ($m)	-100	-200	150	250	200	150	70	30		
	[D] NPV ($m)	326.5									

In this calculation, all net cash flows (CFs) are discounted at an assumed cost of capital of 8%. The impact of the appraisal is a reduction of NPV of 326.5 − 257.8 = $68.7 m, a considerable fraction of the project value. This figure is the combined cost of the appraisal activity plus the loss of NPV due to delay in project start. It does not consider the added value to the project due to the potential to optimise, discussed earlier in Section 5.2.4.

Table 5.5 shows an alternative method of calculating the cost of deferral by discounting the NPV of the development project by the 2 years using the discounting formula and then adding the PV cost of the appraisal activity.

Table 5.5 Components of the reduction in NPV resulting from appraisal.

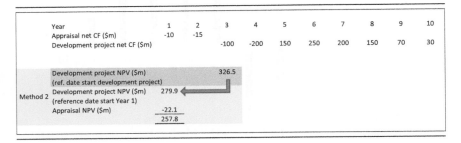

The PV of $326.5 m in Year 3 (2 years deferral) is

$$PV = \frac{326.5}{(1+0.08)^2} = 279.9$$

So the loss of value due to deferral is 326.5−279.9 = $46.6 m.

The PV cost of appraisal is $22.1.

Total reduction in project value due to appraisal activity and deferral = 22.1 + 46.6 = $68.7, as previously calculated, but now separately identifiable.

5.4.4 A medical example of Bayesian revision

This example is included since it is a typical application in medicine to illustrate the impact of a false positive result from a test. Once again the numbers are illustrative but the methodology is robust.

Statistics show that 1% of the population has a particular form of cancer, so the a priori probability of having cancer $P(A) = 0.01$. The a priori probability of not having cancer $P(notA)$ must therefore be 0.99. In the medical world the a priori probability is termed the prevalence of a disease, essentially the reality.

Screening tests for this cancer are understood to be unreliable in that they indicate cancer in 80% of cases when a person actually has cancer, so $P(X|A) = 0.80$. However, screening tests also show a false positive in that they indicate cancer 10% of the time, even when the person does not have cancer, so $P(X|notA) = 0.10$.

Bayes' theorem can be applied to estimate the probability of actually having cancer given the screening test shows cancer $P(A|X)$.

$$P(A|X) = \frac{P(X|A) \cdot P(A)}{P(X|A) \cdot P(A) + P(X|notA) \cdot P(notA)}$$

$$= \frac{0.8 * 0.01}{0.8 * 0.01 + 0.1 * 0.99} = \frac{0.008}{0.008 + 0.1 * 0.99} = \frac{0.008}{0.107}$$

$$= 0.075 = 7.5\%$$

The 7.5% is termed the posterior probability. The probability of the screening test showing positive is the sum of all positive results, the bottom line of the equation (10.7%), but once the new knowledge becomes available that the test is not fully reliable, the posterior probability of actually having cancer, given the test diagnoses the disease is revised to 7.5%.

The equation can be simplified down to

$$P(A|X) = \frac{P(\text{true positive})}{P(\text{true positive}) + P(\text{false positive})}$$

Or even further to

$$P(A|X) = \frac{P(\text{true positive})}{P(\text{any positive})}$$

The following Venn diagram in Fig. 5.20 is an alternative way to calculate the posterior probabilities. The total coloured area contains all possible outcomes (the total event space). Prior probabilities are plotted on the x-axis (not to scale) and the reliability on the y-axis.

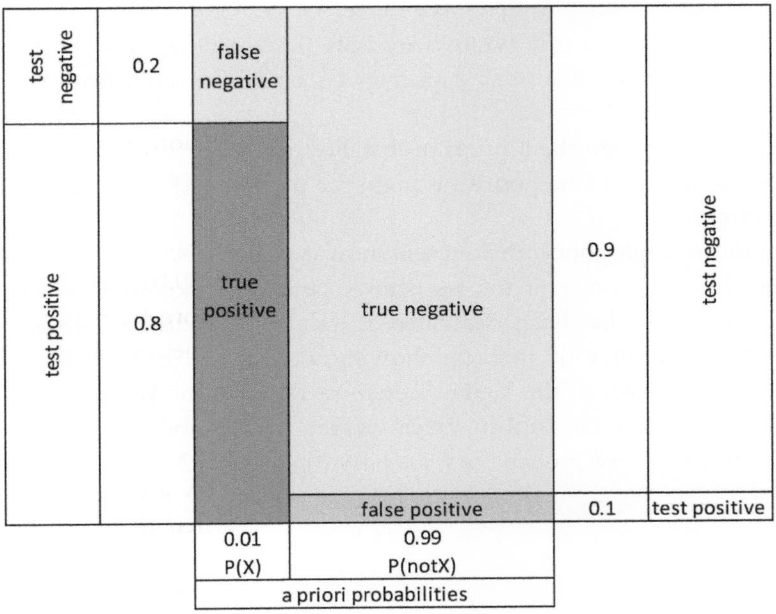

Figure 5.20 Bayes' theorem as a Venn diagram for cancer screening.

In this figure

$$P(A|X) = \frac{P(\text{true positive})}{P(\text{any positive})} = \frac{\text{dark pink area}}{\text{dark pink area} + \text{light pink area}}$$

A further way to approach the application of Bayes' theorem is to use a probability tree, as shown in Fig. 5.21.

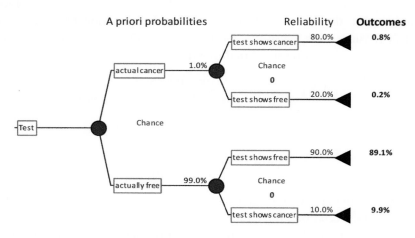

Figure 5.21 Probability tree approach to Bayes' theorem.

In this case all the tests that show cancer are (9.9% + 0.8%) = 10.7% and the tests that show a true positive are only 0.8%, yielding the same result that the probability of having cancer given a positive result from the test is 0.8%/10.7% = 7.5%.

In general, when the a priori probability of a positive result is low and the probability of a false positive is high, the positive test result can be very misleading.

In the example above, the test will show cancer (9.9% + 0.8%) = 10.7% of the time. However, if the test shows cancer the posterior probability of having cancer has been calculated as only 7.5% (actually 7.48%). The revised probabilities of the test showing cancer can be calculated out from the formulae, or the Venn diagram or the probability tree approach.

A very convenient tool in PrecisionTree® is the ability to calculate a Bayesian revision of probabilities, as shown in Fig. 5.22.

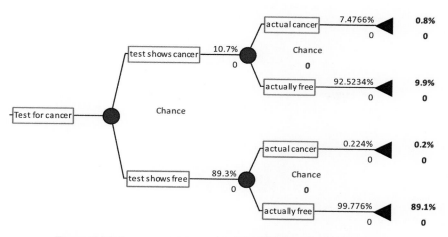

Figure 5.22 Bayesian revision of probabilities using PrecisionTree.

References

[1] R.G. Cooper, Winning at New Products: Creating Value through Innovation, fifth ed., Basic Books, 2017, ISBN 978-0-465-09332-8.
[2] SPE, Petroleum Resource Management System, 2018, ISBN 978-1-61399-660-7. https://www.spe.org.
[3] K. Yates, The Maths of Life and Death, Quercus, 2019, ISBN 978-1-78747-541-0.
[4] The Economist magazine, How Science Goes Wrong, October 21, 2013. https://www.economist.com/leaders/2013/10/21/how-science-goes-wrong.

CHAPTER 6

Project cash flow

Contents

6.1 Cash flow components	133
6.2 Production profile and sales price — the revenue source	136
6.2.1 Oil price assumptions	139
6.2.2 Gas price assumptions	144
6.3 Capex and opex — the technical costs	145
6.3.1 Capex estimation	146
6.3.2 Opex estimation	148
6.3.3 Benchmarks and industry performance on capital costs	152
6.4 Fiscal systems — the host government take	153
6.4.1 Classes and influence of fiscal systems	154
6.4.2 Tax and royalty (concessionary) systems	155
6.4.3 Production sharing contracts/agreements	164
6.4.4 Service contracts	170
6.4.4.1 Pure service contracts	*171*
6.4.4.2 Risk service contracts	*172*
6.4.5 Association contracts	174
6.4.6 Joint ventures	174
6.4.7 Comparison of fiscal systems	175
6.5 Constructing the project cash flow	176
6.5.1 Project cash flow under tax and royalty terms	177
6.5.2 Project cash flow under Production Sharing Contract terms	182
6.6 Discounting — the time value of money	187
6.6.1 Cost of capital	191
6.6.2 Opportunity cost	193
6.6.3 Internal rate of return	194
6.7 Incorporating inflation into the project cash flow	196
References	204

6.1 Cash flow components

Chapter 1 provided an overview of the principles of generating a project cash flow in order to evaluate project economic indicators, and introduced the key terminology used throughout this book. This chapter will provide more detail of the inputs to project cash flow, their relative uncertainties, the construction of the annual net cash flow and its preliminary analysis.

The project scope may range from a major capital project such as a new field development with development costs of multiple billions of dollars, through an incremental investment to enhance an existing project, such as a compressor upgrade or drilling of an additional well ($ millions), down to a small project such as data gathering using a production log ($ thousands). In any case, the project will incur costs and should generate future revenue. In this chapter, it will be assumed that we are considering a major capital project — incremental projects will be addressed in Chapter 9.

The project net cash flow is the balance between the revenues and the costs, usually evaluated on an annual basis as the 'annual net cash flow'. Table 6.1 lists the main components of project costs, and sources of revenue that will be discussed in this chapter.

Table 6.1 Main components of revenue and cost as inputs to project cash flow.

Revenue sources	Cost elements
Sales of hydrocarbon	Capital expenditure (capex)
Tariffs received	Operating expenditure (opex)
	Tariffs incurred
	Government take (royalties and tax)

Gathering the data required to perform a project economic analysis is not a trivial exercise, as it requires discussion with many different groups, each of whom has their own specific knowledge, and, importantly, an awareness of the uncertainty in the data. The petroleum economist will build a 'reference case' project cash flow, followed by an evaluation of the sensitivity of the results to the assumptions made in the inputs. Gathering advice from the specialists on the range of uncertainty in their inputs will become useful for the subsequent sensitivity analysis. Table 6.2 lists the range of specialists to be consulted in gathering the input data, illustrating the broad nature of the task. The uncertainty ranges suggested are relevant to a project being scoped out in the Assess Stage of a project explained in Section 5.1, and may appear rather broad at first sight. However, the ranges are actually realistic given the methods used for their estimation described in this chapter.

Whatever the scale of the project, it requires an investment of capital, generally raised as a combination of bank loans, known as debt, and shareholders' funds, known as equity. The ratio of debt to equity is known as the gearing (but in some cases, gearing is defined as debt divided by debt plus equity). Most companies will raise funding as a corporation rather than for each individual project, using the strength of

Table 6.2 Data required for project cash flow generation.

Data required	Source	Uncertainty range at Assess Stage of project (±% variation from reference case estimate)
Production profile	Subsurface team	±50%
Facilities cost	Facilities engineering	+40%, −30%
Drilling and completion cost	Well engineering	+40%, −30%
Infill drilling, workover cost	Well engineering	±50%
Decommissioning cost	Facilities and well engineering	+50%, −20%
Project schedule	Project management	±20%
Plant operating cost	Operations and maintenance	+40%, −30%
Manpower and overhead costs	Human resources	+40%, −30%
Tariffs	Third party owners	±30%
Fiscal regime	Host government	0% now, highly uncertain in future
Oil, gas, other liquids sales price	Commercial team	±50%
Exchange rates		±30% (country dependent)
Inflation rates		±50% (country dependent)
Cost of capital	Corporate finance	±20%
Country risk rating	Corporate planning	±50% (country dependent)
Social investment obligations		±20%

the corporate balance sheet to achieve a better interest rate on bank loans. A simple objective of any project is to generate a net cash flow which can be used to repay interest on loans, pay down some of the loan capital, pay dividends to shareholders and still have some free net cash flow for re-investment in the same or other projects. Fig. 6.1 illustrates this concept, which will be developed further later in this chapter.

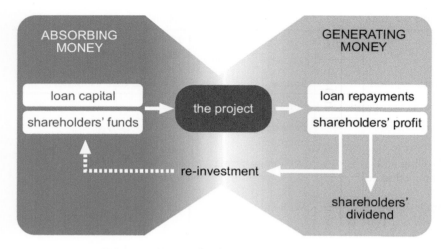

Figure 6.1 Project funding and cash generation.

6.2 Production profile and sales price — the revenue source

The gross revenue earned from a project is the volume of hydrocarbon sold multiplied by the sales price, plus any tariffs received from third parties for use of the processing and export system. The distinction regarding hydrocarbon sales is important, and is not necessarily the same as the produced volume, some of which may be used as fuel, or flared, and these volumes need to be subtracted to calculate the sales volumes.

The subsurface team will provide a range of annual production profiles, which should be checked to ensure that it has accounted for the uptime of the facilities and an estimate of the deduction of fuel and flare volumes. Gas production profiles may be expressed as wet gas, dry sales gas, natural gas liquids (NGLs) and condensate. A wet gas stream implies the total production from the wells, including all components. During separation through the processing plant, the wet gas stream splits into dry sales gas, condensate and NGLs, each of which will have a specific sales price. It is therefore important to ensure that the correct product streams are attributed the appropriate price.

A single estimate of the production profile may be sufficient to create a reference or base case economic evaluation, but a range of forecasts is required to run sensitivities. The conventional range would represent discrete P90, P50 and P10 cases, which represent an 80% confidence range, as discussed in Chapter 4.

Table 6.3 illustrates a production profile for an oil field development with associated gas, using field units. It assumes production export from

Table 6.3 Production profile for an oil field with associated gas sales.

Production year		1	2	3	4	5	6	7	8
Platform oil	Mstb/d	50.0	100.0	100.0	70.0	40.0	20.0	10.0	5.0
Sales oil	Mstb/d	49.50	99.00	99.00	69.30	39.60	19.80	9.90	4.95
Platform wet gas	MMscf/d	20.00	40.00	40.00	28.00	16.00	8.00	4.00	2.00
Fuel	MMscf/d	0.80	0.80	0.80	0.80	0.80	0.80	0.80	0.80
Flare	MMscf/d	0.10	0.10	0.10	0.10	0.10	0.10	0.10	0.10
Export gas	MMscf/d	19.10	39.10	39.10	27.10	15.10	7.10	3.10	1.10
NGL from gas export	Mstb/d	0.096	0.196	0.196	0.136	0.076	0.036	0.016	0.006
Sales gas	MMscf/d	18.15	37.15	37.15	25.75	14.35	6.75	2.95	1.05

Solution GOR	400	scf/stb
Oil stabilisation factor	0.99	Fraction of oil export to sales after pipeline reconciliation
NGL from gas export	5.0	bbl/MMscf
Dry gas yield	0.95	Dry gas/export gas after extracting NGLs

an offshore platform to a terminal facility for final processing prior to sales. After final pipeline reconciliation, the oil exported from the platform leaves 99% available for sales oil. The other 1% in this example is due to final stabilisation of the oil to stock tank conditions (typically 60°F and 1 atm) by removal of light ends.

A portion of the wet gas produced is used for fuel and flare on the platform, and the remaining wet gas is exported to the terminal facility where NGLs are separated to yield dry gas for sales.

The example in Table 6.3 illustrates the corrections that are required to derive accurate sales profiles. The corrections to the platform volumes require the input of process engineers and hydrocarbon accountants who reconcile volumes input into an export system with the sales volumes from the terminal facility. At an early stage of evaluation, this correction may be an unnecessary complication, especially if the final processing facilities are yet to be defined. Platform oil and platform wet gas profiles may be sufficient for a first pass evaluation.

Even in the example, it is debatable whether the small volumes of gas towards the tail end of production would actually be exported — leaving an awkward challenge of what to do with the small volume of gas that is not commercial to export. It may be reinjected into the reservoir, or combined with gas from a later tie-in development, or as a last resort, flared.

Note that for reserves reporting purposes, fuel and flare volumes are generally not accepted as reserves.

The profile presented in Table 6.3 is a single profile, enclosed by a reasonable range of uncertainty above and below the expected profile. The subsurface team should be providing a range of volumes, to be used for sensitivity analysis around a reference case. In the first instance, we will assume that a plateau rate has been estimated, and that the range of profiles would appear as in the left-hand image in Fig. 6.2.

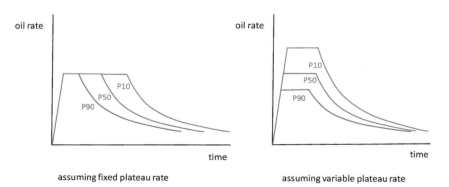

Figure 6.2 Range of production profiles representing subsurface uncertainty.

For a new field development project, the range of uncertainty in the recoverable volumes is often underestimated. The above sketches have a ratio of P10:P90 recoverable volumes of approximately 2. This is realistic given the limited amount of data available at the development planning phase, although this may give an uncomfortable feeling to decision makers and investors. The decision on sizing the facilities and export system should be made by applying petroleum economics to a range of options including designing for the P90, P50 or P10 recoverable volume, and then deciding whether to design for a longer, lower plateau or a shorter, higher plateau, as illustrated in the right hand figure above. Petroleum economics methods provide a consistent tool with which to evaluate these options.

The limited data supporting estimates of subsurface volumes and reservoir production potential create some of the major uncertainties in field development decision-making. Another major uncertainty is the sales price of the hydrocarbons.

6.2.1 Oil price assumptions

Fig. 6.3 shows the oil price over the last century, highlighting the variation since the 1970s.

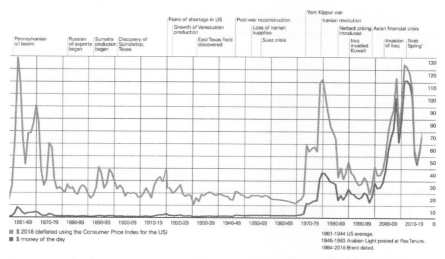

Figure 6.3 History of oil price over the last century. *(Source: BP Statistical Review of World Energy 2019.)*

Most of the abrupt changes in the oil price are a result of change of supply or demand in reaction to geopolitical or global economic events such as war, revolution or shifts in growth. Notable among the events are

- 1973 Yom Kippur war, also known as the Arab-Israeli war
- 1979 Iranian Revolution resulting in the removal of the monarchy
- 1990 Iraq invasion of Kuwait, resulting in a 7-month Iraqi occupation of Kuwait
- 1997 Asian financial crisis and loss of confidence throughout the Asian economies
- 2003 US invasion of Iraq in search of weapons of mass destruction
- 2011 Arab Spring uprisings and rebellions across much of the Islamic world
- 2014 Slowing in Chinese economic growth, influence of growing shale oil production in the United States and sustained production from OPEC member countries

Oil demand is price inelastic, meaning that global demand does not change significantly when price changes. Modest supply changes can then create significant change in the oil price. Between June 2014 and June 2015, oil price more than halved due to oversupply, sending a shock wave through the upstream industry.

It is clear that historic oil price over the last 50 years is highly variable, making its forecasting very challenging. Over the period of a project lifetime, it is likely that the instantaneous oil price will vary considerably. Making forecasts based on the short-term history has severe limitations, and a longer-term view is required for a typical development project whose production lifetime will span several decades.

Various independent bodies make scenario-based forecasts that provide a guide. These use inputs such as forecast growth in world population and the resulting global energy demand, the rise of alternative energy sources as substitutes for oil, oil supply and increase in efficiency of usage to forecast a range of oil price.

For example, the International Energy Agency (IEA) publishes an annual World Energy Outlook (WEO) (Ref. [1]) and the US Energy Information Agency (Ref. [2]) does likewise. Fig. 6.4 is adapted from the IEA World Energy Outlook 2018 and shows a range of oil price forecasts using a scenario-based approach.

The IEA's New Policies Scenario assumes growth in primary energy demand as a result of population growth, urbanisation and economic

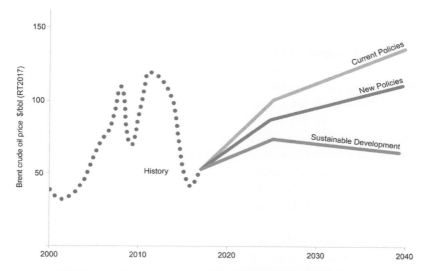

Figure 6.4 IEA forecast of oil price. *(Adapted from WEO 2018.)*

growth, with energy supply strongly influenced by government policies relating to efficiency, renewable resources, curbing of air pollution and phasing out of fossil fuel subsidies. In 2019 this was renamed the Stated Policies Scenario to reflect the inclusion of only specific policy initiatives that have already been announced. The Sustainable Development Scenario assumes almost flat demand to 2040, partly due to increased energy efficiency. The Current Policies Scenario assumes continued growth in the use of the current fuels with little contribution from renewables to meet incremental demand.

The forecasts are clearly diverse and while they represent the IEA's view in 2018, they did not take into account major events such as that of the impact of Covid-19.

Other sources of oil price forecast include Wood Mackenzie's (WoodMac) forecasts (Ref. [3]) which are updated quarterly, or the futures market for oil trading which can be accessed through the International Petroleum Exchange (IPE, Ref. [4]). The IMF publishes its World Economic Outlook in April and October each year, including its assumptions on oil price in short and medium term (Ref. [5]).

Given the significant range of uncertainty on oil price over the producing time of the project, the approach commonly taken by the petroleum economist is to estimate a base case value of the long-term price and then to run broad price sensitivities in the economic model.

For example, a base case may be established using a fixed price of $60/bbl in nominal terms, with sensitivities of $40/bbl and $80/bbl, representing downside and upside outcomes. The phrase nominal terms, also known as money of the day (MOD), implies that the price received is the actual cash in hand, and the oil price has not increased at inflationary levels. A long-term fixed $60/bbl MOD oil price assumption is represented as 'flat MOD $60/bbl'.

An alternative (and one assumed by the IMF for medium-term forecasting) is that the oil price will remain constant in real terms (RT), implying that it increases over time at inflationary levels. Thus, an alternative set of assumptions may be a base case price of flat RT $60/bbl with sensitivities of RT $40/bbl and RT $80/bbl. The use of RT forecasts requires assumptions about inflation levels, and will be expanded upon in Section 6.7.

Fig. 6.5 illustrates the difference between flat RT and flat MOD forecasts, making the point that the flat MOD is a more pessimistic forecast since the oil price is actually decreasing in RT. RT represents the purchasing power of the oil with respect to exchanging it for other goods. For example, if bread costs $1 per loaf now and the price remains flat in RT, its MOD price will increase with the general rate of inflation (GRI). A flat RT $60/bbl oil price means that a barrel of oil will always buy 60 loaves of bread, now and at any point on the future, assuming both commodities rise at the GRI. A flat MOD oil price means that the purchasing power of the oil is declining in RT.

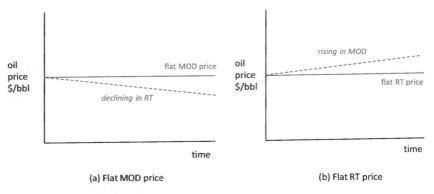

Figure 6.5 Money of the Day (MOD) and Real Terms (RT) oil price assumptions.

The slopes of the lines in Fig. 6.5 represent the annual rate of inflation and are indicative only. If price inflation is compounded then the dotted lines would be curves rather than straight lines. The actual rate of inflation will vary by region but would typically be 2%–5% in western economies. Currently a 3% annual inflation would be considered as typical for these regions.

The oil price forecast represents one of the most uncertain inputs in an economic analysis, and is a matter of taste for each investor. However, within an organisation a common oil price assumption is usually set by the corporate economics function, and applied globally, only to be adjusted according to the relative quality of the crude. If there is an obligation to sell part of the production locally at a pre-determined price, as has been the case in western Siberia, for example, this is an obvious exception.

Benchmark crudes or marker crudes act as a reference point for buying or selling crude oil. The primary three benchmarks are West Texas Intermediate (USA), Brent Blend (Western Europe) and Dubai Crude (Middle East), though others of regional significance include the OPEC Reference Basket (a weighted average quality used by OPEC), Bonny Light (Nigeria), Tapis Crude (Singapore), Urals oil (Russia), Edmonton Par (Canada) and Isthmus (Mexico).

Factors which affect the value of the crude oil are fluid specific gravity, sweetness or sourness (content of sulphur compounds), wax content and transport costs. The American Petroleum Institute introduced the °API measure that relates to the specific gravity of the crude – high API crudes have a higher value as they are more volatile, contain less long-chain hydrocarbons and thus require less energy to crack in refineries.

Brent Crude is a blend of oil from 15 different North Sea oil fields, and although at 38°API gravity it is not quite as light and sweet as WTI crude (39.6°API), it has traded at $2 to $5/bbl higher since 2011, due to changing refinery requirements in the United States, and relative transport costs. Dubai Crude is light (31°API) and sour, containing 2% sulphur and is generally used for benchmarking crude from the Persian Gulf.

In comparison to its regional benchmark, the oil price achieved for each individual field's production may trade at few dollars premium or discount, depending on its chemical and physical properties. Heavier, more sour and waxy crudes will trade at a discount, while lighter sweeter crudes trade at a premium. If the fluid properties for the field being evaluated are well understood, then the oil price adjustment should be included in the assumptions. This raises the importance of representative fluid sampling during the exploration and appraisal phase of the field lifecycle. There are unfortunate examples of designing facilities for sweet oil production, only to later realise that parts of the field contain sour components that require specialised materials of construction, leading to costly retrofits or even limiting further development.

6.2.2 Gas price assumptions

Forecasting gas price raises a rather different set of challenges, as traditionally gas was not considered as a commodity item due to the high cost of transportation. A volume of gas sold locally will be considerably cheaper than the same volume of gas that has been transported over a large distance.

In Chapter 2 it was shown that gas prices varied dramatically depending on geography. In 2018 the average US gas price on the Henry Hub spot market was $3.1/MMBtu, whereas in Japan it was $10.1/MMBtu ($/million British Thermal Units). This is a clear indication that the cost of transportation is highly significant, due to the relative low energy density of gas compared to oil. The gas imported into Japan is in the form of liquefied natural gas (LNG), and includes the cost of the gas, liquefaction, insurance and freight.

However, natural gas is evolving into a global commodity, driven by the surging market in LNG and the continued increase in US shale gas production and export from east coast terminals. In the long term, the geographical price differential for natural gas is likely to reduce.

Gas is often sold on a long-term contract basis, and the price is agreed at the outset of the project between seller and buyer, as discussed in Chapter 2. In this case, the gas price assumption for the economic analysis will reference the terms of the contract.

If gas is to be sold on the spot market, then regional gas price trends may be used to predict the future gas price, and sensitivities run around the base case assumption. Table 6.4 includes examples of regional indices.

Table 6.4 Natural Gas price indices in year 2020.

Region	Index	Reference number
Japan (LNG imports)	Japan Korea Market (JKM)	[6]
Germany	Average German Import Price	[7]
UK	Heren NBP (National Balancing Point)	[8]
Netherlands	Netherlands TTF (Title Transfer Facility) (DA Heren Index)	[8]
US	Henry Hub	[9]
Canada	Alberta	[9]

The TTF and NBP are virtual trading points for natural gas, providing a facility for a number of traders to trade gas futures contracts, physical short-term gas and make exchanges. Henry Hub is a natural gas pipeline located in Louisiana that serves as the official delivery location for futures contracts on the New York Mercantile Exchange (NYMEX). While this is an index based on a centralised location, prices at key shale gas producing basins do diverge from this index.

For uncontracted gas price forecasting, as for oil price forecasting, WoodMac issue a set of forecasts on a quarterly basis (Ref. [3]), which typically make a forecast for eight years. Beyond this, a typical annual inflationary escalation of 2%–3% is commonly assumed. Again, this is only a forecast and sensitivities on the assumptions are recommended in the economic modelling. Alternatively, futures markets may be used as a basis (Ref. [4]), but these are usually run for around five years and beyond this, an assumption on ongoing inflation may be applied.

6.3 Capex and opex — the technical costs

The technical costs involved in a project cash flow comprise capital expenditure (capex) and operating expenditure (opex). These are both investments in the project, but are treated in different ways for management, fiscal and accounting purposes.

Investment in items that are tangible and expected to last for more than a year is typically classified as capex. Examples include wells, platforms, pipelines; all physical items that are literally tangible in the sense that they can be touched. Running costs are classed as opex and include costs of manpower, utilities such as power and water supply, lubricants, tariffs, rental fees and general services which support the daily operations. Capex tends to be front-end loaded in the project schedule as facilities are constructed, and most of the wells are drilled at the beginning of the project development. Opex runs throughout the producing life of the field, and will eventually determine when the economic lifetime occurs — the point at which the revenues from production have reduced to equal the operating expenditure.

The two elements of technical cost are distinguished for several reasons. Firstly, both expenditures need to be controlled for successful project delivery, as initial capex over-expenditure or runaway opex will damage the economics of the project. In order to manage the expenditures, they need to be measured and compared against the project plan. Since they occur at different phases of the project, they need to be measured separately.

Secondly, capex and opex are treated differently for fiscal purposes by the host government. Both the expenditures are treated as 'fiscal costs' or 'tax deductions', meaning that they are allowances against gross revenue in the calculation of taxable income, as briefly introduced in Chapter 1. However, they are treated in a different fiscal manner in that opex allowances are applied in the year of the expenditure, whereas allowances for capex ('capital allowances') are spread over a period of time. The logic for the timing of the capital allowance is that the capex item will have a long useful lifetime, and the timing of the allowance is broadly matched to that lifetime. Fiscal rules in each country will dictate the method of claiming the capital allowance, as addressed in Section 6.4.

Thirdly, the accountants will treat capex and opex items differently in their reporting. Opex will be offset against revenues in the year of expenditure to calculate annual profit. An opex item spent in a particular year will be immediately entered into the profit and loss account as a cost and will therefore reduce annual pre-tax profit, also reducing the annual taxes payable. A capex item will add to the fixed assets of the company in the year of the expenditure and will be depreciated over an assumed lifetime, thus reducing its balance sheet value over time. Simultaneously the annual depreciation will be entered in the profit and loss account as a cost in a similar way to the opex costs. The difference is that opex costs are entered immediately into the profit and loss account but do not appear on the balance sheet, whereas capex costs are deducted over time in the profit and loss account but do appear in the balance sheet as fixed assets. Clearly to account for the capex and opex, which are both part of technical costs, they need to be measured separately.

6.3.1 Capex estimation

Capex estimation is built up by itemising the equipment items, which depend on the facilities concept for development and the type and number of wells. The level of detail available for the items varies with the project stage introduced in Section 5.1.

At the Assess stage of the project, different development concepts will be investigated to determine whether there is at least one feasible development that satisfies the investment criteria, and approximate costs are required to determine this. During the Select stage a more detailed description of the feasible options is required to choose between them. During the Define stage individual process modules will be specified, such as pump type and manufacturer, separator sizes and wall thicknesses, materials of construction for well completion and instrument and control systems, and with this detail more specific cost estimates can be made. During the Execute stage procurement and construction and pre-commissioning is ongoing, and cost control and management of change orders become important, but cost estimates should by now be considerably more accurate, if not fixed through lump-sum contracts with the construction service companies.

Fig. 6.6 illustrates the progression of cost estimating accuracy as the project moves through the stages. It includes reference to the term FEED, Front End Engineering Design, which takes place during the Select stage. Pre-FEED refers to engineering studies performed during the Identify and Assess stages.

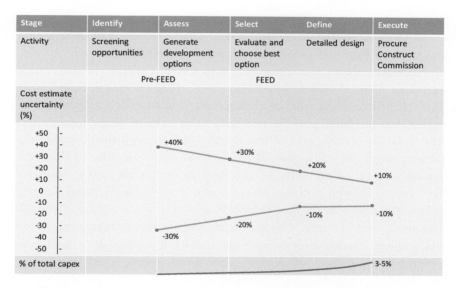

Figure 6.6 Progression of cost estimating accuracy through project stages.

Larger oil and gas companies will develop their own databases for materials and construction costs, but products are also provided by the service sector, such as QUE$TOR from the IHS Group (Ref. [10]). This package is applicable at the Identify and Assess stages for prospect evaluation and generation of development options. The database is suitable for preliminary facilities design and costing, including pipelines, and is updated half-yearly by a research team and through use of feedback from end-users on a regional basis that includes global coverage of offshore and onshore environments.

Inputs required include production profiles (oil, gas, water), numbers of wells, details of export system and a selection of the development concept such as fixed platform, FPSO (floating production, storage and offloading facility), semi-submersible production facility, TLP (tension leg platform) or onshore equivalent facilities. With this input, the components of the production system are sized and costed. Capex estimates include materials, person-hours and associated rates, fabrication, installation, commissioning, certification and contingency costs to account for unspecified cost overruns. Decommissioning costs are also estimated. The cost estimate generated can be considered as a P50 estimate with an uncertainty of ± 25 to 40%, and assumes all costs are in RT, thus excluding any inflationary assumptions. A project schedule is also estimated, producing an investment profile.

Note the skewness of the uncertainty range — there is more scope for cost overrun than cost reduction, compared to the P50 case.

Alternative capex cost estimating systems are available in the market including CostOS from Nomitech (Ref. [11]) and P1 from AGR (Ref. [12]) for drilling and completion cost estimation and scheduling.

Fig. 6.6 also highlights the relatively small fraction of the total project cost spent up to the end of the detailed design performed during the Define stage, typically just 3%—5%. However, by the end of the detailed design, some 90% of the project design decisions will have been made, though nothing has yet been constructed. During detailed design, the potential for cost reduction and optimisation is large, though personnel costs are low. Changes made during the construction period will require change orders, and are inevitably costly. It is useful to remember the mantra of 'plan the execution' during the Define stage and then 'execute the plan' during the Execute stage. The objective is to minimise the amount of changes required during the construction phase of the project. Anyone who has built a house will understand the implications of asking the builder to make changes during construction — this may be a challenging physical task, but almost certainly an opportunity for the builder to add some margin to their profit.

6.3.2 Opex estimation

Operating expenditure is incurred throughout the producing life of the field and can be of similar magnitude to the total capex on an undiscounted cumulative basis (ignoring the timing of the expenditure). It is therefore highly significant. In many respects, opex is harder to estimate than capex which is mostly incurred at the beginning of the project, close to the time of evaluating the development project economics. Forecasting expenditure further into the future carries inflationary and currency exchange uncertainty. Using a software tool such as QUE$TOR provides a bottom-up estimate of opex, based on estimating personnel costs, facilities maintenance and workover activity, but does not include inflationary or exchange rate assumptions.

Because of the long-term nature of the opex, the uncertainty range in opex is similar to that of capex, and it is recommended to run similar levels of sensitivity on opex as was done for capex. So during the Assess stage a very reasonable range of uncertainty on opex is $+40\%$ and -30%, as for capex.

A common method of estimating opex is to base it on fixed and variable cost elements, using regional factors, shown as A and B below:

$$\text{annual opex } (\$/\text{year}) = \text{fixed opex} + \text{variable opex}$$
$$\text{annual opex } (\$/\text{year}) = A \times \text{cumulative capex to date}(\$)$$
$$+ B \times \text{annual production (bbl/year)}$$

The A and B factors will vary widely according to the geographic region and the complexity of the project facilities, but will be in the range $A = 2-5\%$ and $B = 2-10$ \$/bbl. Regional experience and use of analogue producing fields for benchmarking is useful to determine appropriate values for A and B. Note that the fixed opex relates to cumulative capex to date, so that once the initial capex is spent, the fixed opex remains constant until further capex is invested in incremental activities such as plant upgrades and infill wells.

The variable opex is commonly related to the hydrocarbon production. One consequence of this approach is that as production declines over time, the predicted opex also declines. This is often not the experience when producing mature fields, for which annual opex actually tends to increase due to ageing plant suffering from corrosion, erosion, production of hostile fluids and solids, requirements for increased water production and injection and perhaps lack of continuous maintenance. This potentially misleading estimate of opex is particularly true for oil fields where there is a long oil production decline period and increasing water production, with associated increase in water injection volumes required if pressure maintenance is part of the reservoir management plan. Much effort is required to restrict an increase in absolute annual opex, which represents an escalating opex per barrel of oil.

A refined approach to estimation of the longer-term opex for oil fields is to base the annual production on gross liquids (oil plus water) rather than oil production alone. On the assumption that gross liquids remain constant over the producing life of a water-flooded oil field, this will at least predict a constant late-life opex. In fact, for many water-flooded oil fields gross liquid production increases with time, making the assumption of declining opex with time even less reliable.

One common element of opex is tariff payments for use of third party facilities or services, such as pipeline transportation costs, processing of production tied back to a third party host facility or rentals paid to land owners across whose territory a pipeline passes. Such tariffs usually relate to the throughput and are therefore included in the variable opex estimate.

The tariff rate is a negotiated agreement with the third party, who will estimate the client's next best alternative to establish how to approach the negotiation. For example, if the alternative to using the third party pipeline is for the developer to build a dedicated pipeline at a unit cost of say $5/bbl, then the third party can stretch the negotiation close to that figure, say $4.50/bbl. The Forties pipeline system in the North Sea has been host to multiple users other than the pipeline owner, and tariff rates along that pipeline vary considerably depending on the users next best alternative, and incidentally also the oil quality. The pipeline was constructed by BP in mid-1970s, in the very early stages of oil field development in the Central North Sea. It was over-sized for the Forties Field requirements in the reasonable knowledge that it would be available for many later field developments to tie into, and indeed the asset proved to be one of the most profitable single investments made in the region.

In estimating the tariff levels for a new field development, an early approach to the host or a review of analogous tariffs will provide a reasonable base case estimate, around which sensitivities in the modelling can be run.

Whereas capex tends to be front-end loaded in the project investment schedule, opex continues through the producing life, and eventually will determine when to cease production for economic reasons. At the margin, assuming no late-life capex, the economic lifetime is established by the economic limit test (ELT), comparing revenue with expenditure. At the limit, or margin, revenue equals expenditure. The equation below assumes that opex includes all transportation, tariffs, overheads and administration costs.

$$\text{revenue} = \text{expenditure}$$

$$\text{product price}(\$/\text{bbl}) \times \text{production}(\text{bbl}/\text{d}) = \text{opex}$$
$$+ \text{ royalty (if applicable)}(\$/\text{d})$$

Production in this equation is measured in barrels per day, but in the first pass of the economic model, the economic lifetime will be determined based on annual production and annual opex to indicate the year in which the economic limit occurs. Once this date approaches in future, the calculation will be refined to a shorter time period, typically monthly.

Towards the end of field life, production is declining, and opex is flat, or more likely increasing, as indicated in Fig. 6.7.

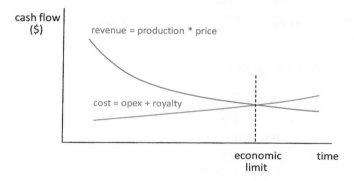

Figure 6.7 Revenue and opex trends in late field life.

Beyond the economic limit, net cash flow will be negative and decommissioning of the project plant and equipment is the logical next step. Chapter 9 will address considerations that may postpone decommissioning, including the removal of royalty, reduction of opex and potential for increasing production.

The economic limit is not the only factor that can determine the end of production of the field as far as the owners are concerned. A lack of mechanical integrity of the facilities, often a health, safety or environmental (HSE) consideration, may give rise to a Cessation of Production (COP) date. Alternatively, a production licence may expire and the ownership of the field may transfer to a new owner or to the host government in a production sharing agreement (PSA). In terms of reserves reporting, the earliest of the economic lifetime, as determined by the ELT, the COP date or licence expiry determines the original owners' reported reserves.

Netback price is a useful benchmark for comparing the efficiency of production. It does not have a universally specific definition, so caution is required in its interpretation, but a common definition is

$$\text{netback} = \text{revenue} - \text{royalty} - \text{opex} - \text{transportion costs}$$

Care is required to ensure that the opex above does not already include transportation costs. Netback price is often expressed on a dollar per barrel or dollar per barrel of oil equivalent (boe) basis and netback/boe becomes a useful metric for measuring efficiency of production over time, or comparing companies to the competition.

A useful reference point for the start of transportation is at the wellhead, in which case the term applies on a dollar per boe basis to become the 'netback price at the wellhead'. In this case, the economic lifetime can be considered when the netback price/boe equals the current sales price of the product.

6.3.3 Benchmarks and industry performance on capital costs

Benchmarks for capex and opex include the Upstream Capital Costs Index (UCCI) and Upstream Operating Costs Index (UOCI) from IHS Markit (Ref. [13]) which provides predictions of short-term movements and long-term trends in costs. The Rushmore Drilling Index (RDI) is a useful source of cost estimation for well costs and schedule (Ref. [14]).

McKinsey, a consultancy group, report that in a study of 800 major capital projects, with investments exceeding $1 billion, on average projects were one year behind schedule and 30% over budget, and expect the same to be true of smaller projects, even down to the $10–$20 million level (Ref. [15]). Capital cost overrun and delay have a dramatic impact on the base case economics, as will be demonstrated in Section 8.6 (sensitivity analysis). Conversely, identifying a realistic opportunity to accelerate a project schedule or reduce costs can create an upside value compared to the base case.

In 2014 Ernst and Young (EY research) analysed 365 complex projects and revealed that 64% of projects faced cost overruns and 73% reported delays. Completion costs were, on average, 59% higher than initial cost estimates (Ref. [16]). This adds to the overwhelming evidence that there is significant uncertainty in capex estimation, even at the end of the Define stage when the Final Investment Decision (FID, also commonly known as Sanction) is made.

Rystad Energy, a research and consultancy company, estimate the breakeven price per barrel of Brent Crude in key oil-producing regions in 2019 as shown in Table 6.5 (Ref. [17]). This is equivalent to the total capex and opex for development of new fields.

Table 6.5 Breakeven oil prices by region, 2019 (Rystad Energy).

Region	Weighted average breakeven oil price ($/bbl)
Onshore Middle East	42
North America tight oil	46
Shelf	49
Deepwater	58
Russia onshore	59
Extra heavy oil	59
Rest of world onshore	60
Oil sands	83

North American tight oil development and production costs reduced dramatically during 2015—20, to the point where their breakeven cost is within a whisker of onshore Middle East. The United States was the largest oil-producing country in the world during 2019, with production averaging 12.9 million barrels per day; such has been the upswing in tight oil production from the Permian Basin.

Benchmark figures are useful in the Identify and Assess (pre-FEED) stages of economic evaluation. If the technical cost per barrel is broadly in line with the regional metric it confirms that the appropriate development concepts are being considered.

6.4 Fiscal systems — the host government take

Petroleum fiscal systems are arrangements for sharing the net revenue from the production of hydrocarbons between the host nation and the operating companies, net revenue being the difference between the value of the product and the cost of production. The reason that the petroleum industry has its specific fiscal systems, rather than just application of straightforward corporation taxes (CTs) as applied to most other industries, is that the net revenue is assumed to be extraordinarily high. When the United Kingdom introduced specific taxes for the North Sea petroleum industry in the early 1970s, it was on the basis that the operators were making 'excess profits' due to the net revenue, such that the standard CT would leave the oil companies with too high after-tax profit.

The fiscal system is distinct from the petroleum licencing system; the latter grants the rights to exploration and production to commercial entities, meaning oil companies, partnerships or consortia who act as the operator. The fiscal system determines how governments and operators share the net revenue.

The term fiscal derives from the Latin 'fiscus', meaning the roman emperor's purse, in other words the public treasury that collected taxes. John Maynard Keynes (1883—1946), a British economist who developed theories known as Keynesian economics, brought the term into common use.

Modelling the fiscal terms is the most challenging aspect of building an economic model, and the impact of the terms on the attractiveness of the project is very high. The contractor's share of profit from a barrel of oil produced, known as the contractor's 'take', can vary from 10% to 75%, depending on the fiscal system. There are more fiscal systems than there are countries. Within one country, different terms may apply to different fields since each set of terms is negotiated separately, and even within a field lifetime, the fiscal terms may change. In the cases where fiscal terms are negotiated, the discussions can be fractious and protracted, since the interests of the host government and the contractor are opposed in that each wants to maximise its share of the net revenue. Interests are, however, aligned in both wanting the field to be developed in a way that maximises the net revenue, while meeting HSE and legislative standards. A final agreement implies that both parties believe that their required economic, HSE and social metrics will be achieved under the agreed terms.

The subject of fiscal systems is complex and dynamic. The fiscal terms can make or break a project, and are as important an element of the economic evaluation as the geology, geography or country risks. This chapter will illustrate examples of fiscal terms, and refer to key sources of information with which to keep updated on the fiscal terms.

The chapter will focus on concessionary and contractual agreements, being the two major types of fiscal terms, and uses the term 'contractor' in contractual systems and 'operator' in concession systems, though both terms represent the same commercial entity.

6.4.1 Classes and influence of fiscal systems

The jargon used in describing fiscal systems can be confusing, and is often used loosely. In this book, the fiscal system will encompass concessionary systems and contractual systems. The distinction between these is as follows.

In a concessionary system, such as the United Kingdom, the contractor takes ownership of the production and pays royalty on the gross revenues from production, and also pays tax on the profit element of the cash flow. This is often known as a 'tax and royalty system'. In principle, royalty can be paid in cash or in kind, the latter meaning payment through the actual volume of oil produced.

In a contractual system, the ownership of the production (or 'title' to the production) remains with the host government. The oil company finances and carries out all E&P operations and is considered as a contractor, providing technical services to the government. Compensation is received as a share of the oil or gas production for the recovery of its costs, along with a share of the profits, also paid in the form of oil or gas.

Host governments are represented by the national oil company (NOC) or by the oil ministry, and may choose to take part in the investment as part of a consortium or joint venture.

In a service contract or risk service contract, the contractor again finances and operates the project and receives a fee for this service, which can be in cash or in kind. The fees typically allow the recovery of all or part of the contractor's costs plus a form of profit component.

Fig. 6.8 shows the key petroleum fiscal systems, and is adapted from a seminal book by Daniel Johnston (Ref. [18]). Additional sources of updated terms of global fiscal systems can be found in annual publications by Ernst & Young and by Deloitte (Refs. [19—20]).

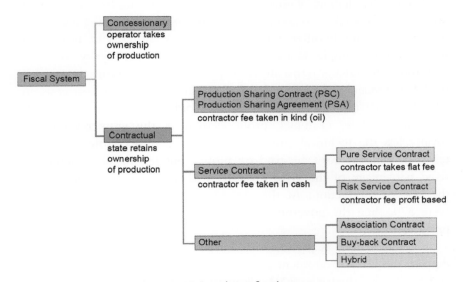

Figure 6.8 Petroleum fiscal systems.

6.4.2 Tax and royalty (concessionary) systems

Approximately half of the world's fiscal systems are concessionary, in which the government take is based on a tax and royalty. The ownership of the production, also known as the 'title', rests with the oil and gas company upon production. The government takes an initial share of the production in the form of royalty, which is one of the oldest forms of government take, based on the concept that mineral rights belong to the crown, therefore 'royal' minerals can only be mined if a payment is made to the crown. Tax and royalty is the more conventional form of government take and is prevalent in countries that commenced oil and gas production in the early stages of the industry, such as the United States, Canada and Western Europe. Contractual systems started to be introduced in the 1960s, with Indonesia leading the way.

Royalty may be paid in cash; a fraction of the sales price, or in kind; a fraction of the oil or gas production. When royalty for oil and gas production was introduced in the UKCS (United Kingdom Continental Shelf) in 1975, the rate was 12.5%. Operators could pay this in cash, or in kind, but either way, the government take is immediate, with no allowances for the cost of production. So, in terms of the cash flow

$$\text{royalty (\$)} = \text{gross revenues (\$)} \times \text{royalty rate (\%)}$$

The royalty element of government take is a regressive form of taxation. Since the royalty takes no account of the cost of production as a deduction, it becomes more punitive for the operator as the cost of production increases. For example, at a per barrel oil price of $80, royalty would be $10 under the UKCS terms above. If per barrel production costs are $20 then $50 remains as a profit element. If per barrel production costs are $40 then $10 is still paid as royalty and only $30 remains as a profit element. In these examples, profit element is the gross revenue less royalty less the technical cost of production, a rather loose use of the term. In general, taxes are paid on profits, but the royalty scheme is a fixed percentage of the gross revenue, with no deductions allowed for the cost of production. At the extreme, if per barrel production costs were $70, then the government take is still $10 as royalty, leaving no profit element.

For this reason, as production costs rose in the UKCS and oil price dipped, the UK Government progressively phased out royalty from 1983 and finally abolished it in 2003, allowing a reasonable profit margin upon which taxes were paid, but leaving an acceptable margin for the operator. This led to more enthusiastic investment in field developments by operators, also extending the field life of mature fields, as will be discussed in Chapter 8.

In some concessionary schemes where royalty applies, the royalty rate is linked to production or gross revenue on a sliding scale, increasing by incremental steps. In all cases, royalty will be payable from the start of production.

Unlike royalty, taxation is a progressive mechanism in that certain deductions are offset against gross revenue to calculate taxable income, to which the tax rate is applied to calculate the tax payable. The typical tax deductions include royalty, opex and an allowance for capex, termed 'capital allowance'. Tax deductions are also known as fiscal costs or fiscal allowances.

$$\text{tax}(\$) = \text{taxable income}(\$) \times \text{tax rate}(\%)$$
$$\text{taxable income}(\$) = \text{gross revenue} - \text{tax deductions}(\$)$$
$$\text{tax deductions}(\$) = \text{royalty} + \text{opex} + \text{capital allowance}(\$)$$

Tax is calculated on an annual or half-annual basis, but for convenience, this chapter will assume annual payments. Firstly, the tax deductions must be assessed to allow the taxable income to be computed, and then the tax can be calculated. There may be several levels of taxation, and the first form of tax is usually deductible prior to calculating the next level. More of this later in the chapter, but in a simple example, the revenue from the production of a barrel of oil under a tax and royalty system is illustrated in Fig. 6.9, using a royalty rate of 12.5% and a tax rate of 50%.

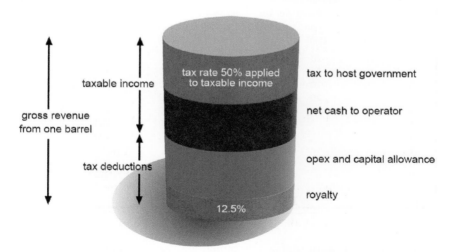

Figure 6.9 Split of revenue under a tax and royalty system.

Royalty is paid in the first year that production commences and continues to be paid as long as the field produces. Opex is claimed as part of the tax deductions in the year of expenditure, whereas capex is claimed as a series of tax deductions, spread out over a number of years as what is termed capital allowance. The basis for this is that the operator has invested in a capital item, the lifetime of which is greater than a year. The mechanism for spreading out the capital allowance will be prescribed in the fiscal terms, but is usually one of three forms, as follows.

The **straight line capital allowance method** spreads the capex out over a prescribed number of years in equal quantities. The **declining balance method** allows the operator to claim a fixed percentage of the unrecovered capex each year. In the example shown in Table 6.6 both methods use a rate of 20%, and assume an economic lifetime of 10 years. The comparison shows that the declining balance method spreads the allowances over a longer period, and may also leave a tranche of unrecovered capex at the end of the project, which will be recovered 100% in that final year — a rather inelegant profile of capital allowance. In the case of negative taxable income in the final year, it could happen that this unrecovered capex is never recovered.

Table 6.6 Straight line and declining balance methods for capital allowance (both 20% p.a.).

Year	Capex	Straight line capital allowance 20% p.a.				Declining balance 20%	
		Allowance for Year 1 Capex	Allowance for Year 2 Capex	Allowance for Year 3 Capex	Total capital allowance	Unrecovered capex at year end	Capital allowance
	($m)	($m)	($m)	($m)	($m)	($m)	($m)
1	100	20			20	100	20
2	400	20	80		100	480	96
3	200	20	80	40	140	584	117
4		20	80	40	140	467	93
5		20	80	40	140	374	75
6			80	40	120	299	60
7				40	40	239	48
8						191	38
9						153	31
10						122	122
Total	700				700		700

Both methods assume that the capital allowance cannot start until the capex has been spent, but can be claimed as a tax deduction against production as soon as spent — implying that there is a gross revenue against which to claim. This assumption will be challenged when 'ring fencing' is introduced later in this section.

To avoid the issue of the tranche of unrecovered capex at the end of field life, a third approach is the **unit of production method**. This follows the accounting principle of matching the allowance to the usage of the capital item by relating the allowance to the fraction of the remaining recoverable volume of hydrocarbon that is produced each year. Table 6.7 shows the same capex profile as above but now uses the annual total production and remaining recoverable volume (reserves) at the start of each year to calculate the annual depletion factor, being the annual production divided by the remaining recoverable volume at the start of the year. This depletion factor is then multiplied by the unrecovered capex at year-end to calculate the annual capital allowance. Now the capital allowance fully recovers the capex by the end of field life.

Table 6.7 Unit of production method for capital allowance.

Year	Capex	Annual production	Unit of production method			
			Remaining reserves at year start	Depletion factor	Unrecovered capex at year end	Capital allowance
	($m)	(MMbbl)	(MMbbl)		($m)	($m)
1	100	25	250	0.10	100	10
2	400	40	225	0.18	490	87
3	200	50	185	0.27	603	163
4		50	135	0.37	440	163
5		40	85	0.47	277	130
6		25	45	0.56	147	81
7		15	20	0.75	65	49
8		5	5	1.00	16	16
9						
10						
Total	700	250				700

While this appears to be an elegant method to match the capital allowance to the fractional use of the facility over the economic lifetime of the project, it assumes that the estimate of ultimate recovery from the field remains unchanged. This is hardly ever the case in practice — as new data and technology become available the estimate of ultimate recovery and therefore remaining reserves will change, and this will give rise to step changes in the capital allowance profile.

The method applied to calculate the capital allowance profile is not a matter of choice for the operator — it will be dictated by the fiscal terms set by the host government. These need to be understood when creating the economic model, and care is required, since different capital items may have different rates of decline applied. At the extremes, a pipeline may have a decline rate of just 10% per annum, while an individual well may have a 100% decline rate, meaning that although it is classed as a capex item, the 100% p.a. decline rate effectively means that it is treated as opex. This is an example of a fiscal term which incentivises expenditure on wells, since the faster the capital allowances are recovered, the less taxable income and thus less tax is paid in the early years of the project. Because of the discounting methods applied to cash flows, introduced briefly in Chapter 1, later costs such as tax payments favour the NPV of the project.

The examples shown in Tables 6.6 and 6.7 assumed that as soon as the capex is spent, capital allowance can be claimed. This is valid as long as there is production from the field in Year 1. However, if there is no production until, say, Year 3 then the capital allowance cannot start to be claimed until Year 3 as it is deductible against gross revenue. In such cases, the capital allowance is accrued and claimed once production starts, as shown in Tables 6.8 and 6.9. If in any year the capital allowance exceeds the cash flow against which it can be claimed (gross revenue - royalty - opex), then the unrecovered amount must be carried forward to the next year.

Table 6.8 Straight line and declining balance capital allowances with deferred production.

Year	Capex	Straight line capital allowance 20% p.a.				Declining balance 20%	
		Allowance for Year 1 Capex	Allowance for Year 2 Capex	Allowance for Year 3 Capex	Total capital allowance	Unrecovered capex at year end	Capital allowance
	($m)	($m)	($m)	($m)	($m)	($m)	($m)
1	100					100	
2	400					500	
3*	200	20	80	40	140	700	140
4		20	80	40	140	560	112
5		20	80	40	140	448	90
6		20	80	40	140	358	72
7		20	80	40	140	287	57
8						229	46
9						184	37
10						147	147
*first oil							
Total	700				700		700

Table 6.9 Unit of production capital allowance with deferred production.

Year	Capex ($m)	Annual production (MMbbl)	Unit of production method			
			Remaining reserves at year start (MMbbl)	Depletion factor	Unrecovered capex at year end ($m)	Capital allowance ($m)
1	100				100	
2	400				500	
3*	200	25	250	0.10	700	70
4		40	225	0.18	630	112
5		50	185	0.27	518	140
6		50	135	0.37	378	140
7		40	85	0.47	238	112
8		25	45	0.56	126	70
9		15	20	0.75	56	42
10		5	5	1.00	14	14
* first oil						
Total	700	250				700

The procedure of claiming tax deductions (royalty, opex and capital allowance) against gross revenues raises the subject of 'ring fencing', which has been introduced into petroleum contracts or concession systems to limit the extent to which allowances can be claimed against revenues. The ring fence is a boundary that defines at what level each allowance (e.g. royalty, capital allowance and tax) can be claimed. The boundary may be at a field, licence block or concession level, or even at an entity level, such as upstream and downstream businesses.

For example, if the capital allowance is ring fenced at field level, the allowance can only be claimed against revenue from within the ring fence, and would have to be deferred until production commences from the field, as in Tables 6.8 and 6.9. If the capital allowance is non-ring fenced and can be claimed from production of another field then the capital allowance may be claimed in the year of expenditure, as in Table 6.6, even if production from the field being evaluated does not start production until later. The ring fence may vary for different forms of allowance. For example, a special petroleum tax may be an allowance against CT within a field ring fence, but CT may be ring fenced at corporate entity or country level. Onshore USA a State Tax is charged and paid to the State, which then becomes a tax deduction against revenue when computing Federal Tax, paid to the national government.

This hierarchy of fiscal calculations is complex, and is sometimes overlooked in calculating a project cash flow, but requires attention to detail, as described by D. Smith (Ref. [21]).

If a field development plan has a ring fence around it for taxation purposes, then tax deductions within that ring fence will be deferred until there is sufficient revenue against which to claim them, generally only commencing at first production date. This is a similar situation to an operator starting its first field development in the country — a term known as a 'newcomer', who will have to wait for first production from the field before being able to claim the tax deductions. This can significantly disadvantage the newcomer compared to an existing taxpayer, since the newcomer's tax deductions are deferred. In this regard, a newcomer is in a similar situation to being ring fenced, as illustrated in Fig. 6.10.

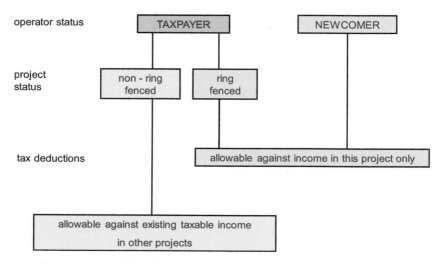

Figure 6.10 Similarity between ring-fenced field and newcomer status.

A newcomer to a region can improve their fiscal position by farming into an existing producing field, and using the revenue from their share of production from the existing field to claim the non-ring-fenced tax deductions.

In the early days of production in the North Sea, newcomers were offered a share of the producing Forties Field, operated and largely owned by BP at the time. A small share (0.25%) was sufficient to provide revenues against which to offset exploration and appraisal costs against certain elements of the tax payable on Forties revenues, thus reducing the after-tax cost of E&A activities. BP was able to sell the small share at a premium, advantaging both companies. The UK government later closed such opportunity by introducing a Ring Fence Corporation Tax (RFCT), meaning that only costs incurred within the field's ring fence were eligible deductions. This is an example of a changing set of fiscal conditions during the lifetime of a field development under a concessionary system.

To reinforce the variability of fiscal terms under a concessionary system, Fig. 6.11 illustrates some of the key changes made by the UK government for oil and gas fields since the commencement of oil production in the 1970s. The changes are generally in response to oil price fluctuations, becoming more punitive at high oil price and relaxing as oil price eases, and introduced as the Treasury announces a new budget statement.

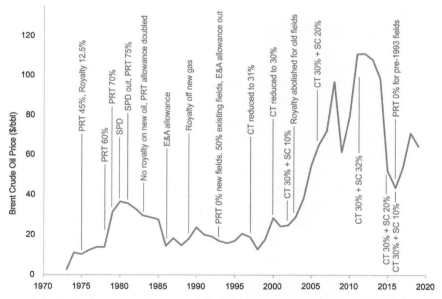

Figure 6.11 Variation of fiscal terms for oil and gas production in UKCS 1975—2019.

The data for 1973—75 are OPEC crude oil prices, and Brent Crude oil prices from 1976, from which point it became a regional marker. The many changes to the fiscal terms are the UK government's actions to try to maximise revenue for the State, while still providing a sufficient incentive for continued investment in the UKCS. Many varieties of government take are evident, including Royalty, Petroleum Revenue Tax (PRT), Special Petroleum Duty (SPD), CT and Supplementary Charge (SC). At the peak of these charges, the incremental government take was 83.5%.

A consequence of the fluctuating fiscal system in the United Kingdom is that operators have difficulty in anticipating the next change, and can be reluctant to invest in an unpredictable fiscal environment such as this. By contrast, in the same offshore region, the Norwegian taxation system has been high at an incremental government take of 78%, but has remained constant since 1990, providing operators predictable fiscal terms. While this Norwegian oil and gas tax rate seems punitive, tax deductions are generous, including offset of exploration costs against all taxes for taxpayers, and reimbursement of 78% of exploration costs for newcomers. This is also a reminder that the headline government take is not a pure indicator of the harshness of the terms — consideration of all fiscal aspects is required.

6.4.3 Production sharing contracts/agreements

Production sharing contracts (PSCs) were first introduced in Indonesia in the 1960s as an agreement between the Indonesian Government and international oil companies (IOCs) for petroleum development. The Government maintained the sovereignty to the petroleum resource, sometimes called the 'title'. The oil company is referred to as the contractor, and the NOC, such as Pertamina in Indonesia (Permina in the 1960s), may represent the government. The term PSC is synonymous with PSA.

Under a PSC the contractor pays for all E&A and development costs and recovers these costs from a share of the production if the project becomes a commercial development. This mechanism, known as 'cost recovery', allows costs to be recovered from a specific fraction of the production, called 'cost oil'. E&A costs are usually, but not always, recoverable from cost oil.

Some PSCs include a royalty payment, which is taken directly by the government as a fraction of production. This may be as high as 15%, but note that many PSCs do not include a royalty element. After deducting royalty and cost oil, the remaining fraction of production is termed 'profit oil' and is shared between the contractor and the government, hence the term production sharing. The contractor then pays tax on its share of the profit oil. Fig. 6.12 illustrates a PSC in which royalty is charged at 10%, the maximum provision for cost oil is 50%, and the profit oil is shared 50/50 between the contractor and the government, with the contractor then paying the government tax on its share of profit oil, illustrated as 40%.

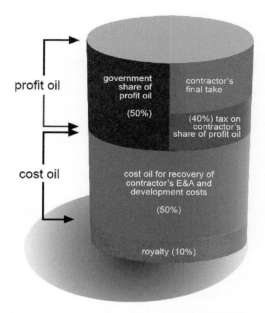

Figure 6.12 Split of production under a particular production sharing contract.

Assuming all of the cost oil is required to recover costs, the contractor's final take from this agreement with a 40% tax rate on the contractor's share of profit oil is

$$(1 - \text{royalty} - \text{cost oil}) \times (1 - \text{government share of profit oil})$$
$$\times (1 - \text{tax rate}) = (1 - 0.1 - 0.5) \times (1 - 0.5) \times (1 - 0.4)$$
$$= 0.12 \text{ or } 12\%$$

It should be realised that over the field lifetime, the contractor will recover the costs through cost oil, but this is just neutral in terms of net take. In reserves reporting, the contractor may report the cost oil plus the contractor's share of the profit oil. This represents the contractor's entitlement. If the contractor is a consortium then each member of the consortium must multiply the entitlement by their working interest to calculate its reserves.

The cost oil is a maximum fraction of the production that may be claimed, 50% in the above example. In the early years of production, there will be insufficient cost oil available to recover both the early opex and the initial capex of the project. The maximum allowable cost oil will be claimed and any unrecovered costs will need to be carried forward to subsequent years. In later years of the field life, when there is little capex, and the opex spent is less than the 50% allowable, then the opex spent is claimed, leaving an increased fraction of profit oil to be shared between the contractor and the government. In other words, the 50% should be viewed as a ceiling or cost oil maximum (COM) and the contractor takes the lower of either the COM or the actual annual expenditure.

There is a common incentive for contractor and government to control project costs, since less cost means less cost oil recovered, leaving more profit oil to be shared. However, this is not always the perception within the contractor's office, as costs are often considered to be fully recoverable against cost oil and therefore perceived not to need careful management. This is a misconception; the lower the costs, the more profit oil remains, to the benefit of both contractor and government. Not surprisingly, the government, often represented by the NOC, take a keen interest in cost control, providing a useful check on both capex- and opex-related activities.

With the exception of the tax rate, which is usually the country's CT rate, all of the percentages shown in Fig. 6.12 are negotiated between the contractor and the government, for each field, block or contract area. In the first instance, outline terms may be provided to the contractor, but are open to discussion. If the outline terms make the project economics unattractive to the contractor, a negotiation is likely ensue. This can be prolonged and intense, as the company strives to realise terms that meet the investment criteria while the government wants to maximise recovery from the field to protect its national interest. The PSC discussions for the In Salah gas project in Algeria continued between BP and Statoil as the IOCs and the NOC Sonatrach for over 3 years before reaching a final agreement. Elmaci reports that the negotiations between Italian contractor ENI and the Kazakhstan government for the PSC terms of the Kashagan oil field development took 23 years; exceptional due to the size, technical complexity, political and financial issues (Ref. [22]). These protracted negotiations will damage the lifecycle project economics as they extend the time between first investment in gaining entry and the first revenue from production.

In addition to the split of the production, the PSC terms include a licence period for which the contract is valid. Fig. 6.13 illustrates the typical time periods defined in a PSC. The upper half of the figure shows a project running to the agreed schedule, while the lower shows some delay in the E&A activities, giving rise to a portion of the production that runs beyond the PSC production period.

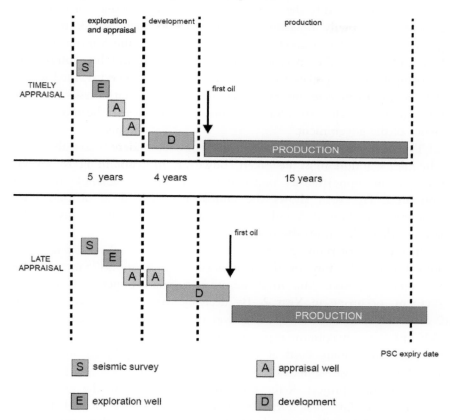

Figure 6.13 Typical periods defined in a PSC showing impact of delay.

The time allowed under the PSC terms for E&A, development and production in Fig. 6.13 is 5, 4 and 15 years, respectively. If the first oil date is delayed, the planned 15 year production profile may run beyond the end of the production period of the PSC, at which point the PSC expires. In principle, all plant and equipment associated with the project becomes the property of the government, since all of the contractor's costs have been recovered through cost oil. The government then has the right to take over the equipment and continue to operate the field, usually passing this responsibility to its representative, the NOC. This is of course detrimental to the contractor since 100% of the production beyond the end of the PSC belongs to the government.

With the prospect of losing its share of the tail end of the production profile, the contractor has some options. Firstly production could be accelerated compared to the base profile. However, the government usually approves a quarterly operations and production plan, and if there are grounds to expect that acceleration may damage the ultimate recovery of hydrocarbons then the government may reject this proposal. Secondly, the oil company can negotiate to extend the production period defined in the PSC, and in many cases this proves successful, though not necessarily under the same terms. The terms for the contract extension may be modified in favour of the government.

The best course of action for the contractor is to meet the PSC schedule. Caution must be taken, however, in committing to several simultaneous opportunities under PSC terms if the schedule is to be honoured. Fig. 6.14 shows the annual net cash flow of two PSC opportunities (Projects A and B) being carried out simultaneously on the left and then phased on the right.

In the case of running the projects simultaneously, the most negative annual net cash flow of the combined projects is -$475 m and the maximum exposure (the most negative cumulative net cash flow) is -$665 m, occurring in Year 2. By delaying the PSC for Project B by 2 years, this reduces to a most negative combined annual net cash flow of -$275 m and a maximum exposure of -$505 m, in Year 3. Additionally, the decommissioning costs at the end of the two projects are phased, spreading the negative net cash flow in the final years.

For a single project, if the economic life falls within the production period of the PSC, it is normally funded by the contractor and the cost claimed as part of the cost oil. However, at this point there will be no production and hence no cost oil available. If the economic life falls outside the production period of the PSC, the project has now become 100% the property of the government, which then takes the entire liability for the decommissioning activity and associated cost. To avoid these complications, which can create poor relationships between IOC and government, the PSC terms commonly require that a fraction of the annual revenue from production be diverted to a decommissioning fund or account, which will be held in escrow in custody of the government. The account balance accumulates over the producing life of the field, and is released when decommissioning is required. If applied, this element should be included in the economic model.

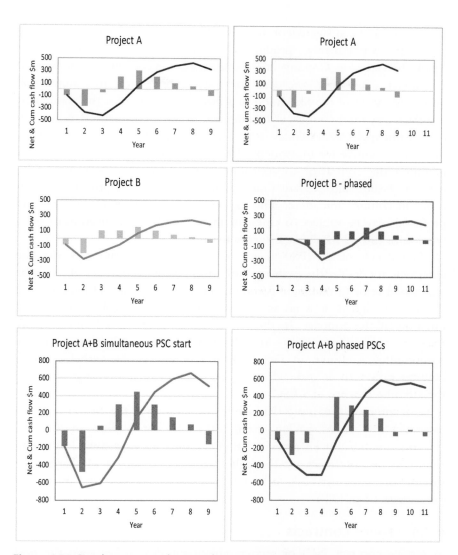

Figure 6.14 Simultaneous vs phasing of contracts under PSC terms; net cash flow and cumulative.

The outcome of contract extension discussions is another of the commercial uncertainties facing the contractor. Competent operatorship and maintenance of good relations with the government play a role in the outcome, along with the desire of the government to take operatorship through the NOC. In terms of reserves reporting, unless a contract extension has been agreed or is deemed highly likely to be agreed, the contactor's entitlement to production beyond the current contract period should be classed as Contingent Resources, contingent upon being granted a contract extension. Precedent for contract extension in-country will inform the view of the likelihood of a positive outcome for the contractor.

PSCs vary from country to country and from field to field. Royalty may or may not apply. Cost oil may extend in some cases to 100% of production, being an incentive for investment. The contract terms may include clauses requiring the contractor to contribute from their share of profit oil to national health and welfare funds, transportation development funds or other in-country initiatives. The export of the contractor's cost oil and profit oil may be subject to tariffs, or there may be an obligation to sell a fraction of this oil locally at a discounted price.

The details of the PSC need to be carefully incorporated into the economic model for the evaluation of an opportunity, and are likely to be negotiated prior to making a commitment to the contract. However, once the PSC is signed, it is an agreement, and the terms will apply for the duration of the contract. In this sense, the PSC provides more certainty than a concession system in which tax and royalty rates change significantly, as shown in Fig. 6.11 in the UKCS. Nevertheless, other country risks such as nationalisation and political regime change may still threaten the security of a PSC.

It is interesting to note that while Indonesia was the first country to introduce PSC terms, having practised this for over 50 years, in 2017 the government introduced a revised model called a Gross Split PSC. This removes the cost recovery through cost oil and instead of sharing the production as profit oil, it shares the gross revenue in an agreed percentage (Ref. [23]).

6.4.4 Service contracts

A service contract (SC) or service agreement (SA) is between the host government and the contractor, and is in many ways similar to a PSC in that the contractor pays for all E&A, development and operating costs which are then recovered from production over time. In addition to this cost recovery, the contractor is paid a fee for their technical services, which is related to the level of production. In the case of a flat fee related to production, say $2/bbl, the contract is known as a pure service contract. If the fee is related to profit (or more accurately the net revenue, the difference between gross revenue and cost of extraction), then it is called a risk service contract. As in the PSC, resource ownership or entitlement rests with the host government.

6.4.4.1 Pure service contracts

The pure service contract guarantees the contractor a specific net cash flow per barrel, and in times of low oil price, this can seem advantageous. However, at high oil price this is not so attractive, as Table 6.10 illustrates. Here the cost recovery is assumed to be $5/bbl to recover capex and opex, and the flat fee is $2/bbl. The oil price assumptions may seem extreme, but prices fell outside this range in the period 1998 to 2008, a time period well within the typical service contract duration. Such service contracts based on a fixed fee have been applied in parts of Nigeria, sometimes with the contractor fee being even less than $2/bbl.

Table 6.10 Contractor and government take under a pure service contract at different oil prices.

	Oil price $20/bbl	Oil price $80/bbl
Gross revenue ($/bbl)	20.0	80.0
Technical costs ($/bbl); reimbursed	−5.0	−5.0
Net revenue ($/bbl)	15.0	75.0
Contractor fee ($/bbl)	−2.0	−2.0
Government take ($/bbl)	13.0	73.0
Contractor share of net revenue	=2.0/15.0 = 13.3%	=2.0/75.0 = 2.7%
Government share of net revenue	=13.0/15.0 = 86.7%	=73.0/75.0 = 97.3%

In this example, the net revenue is defined as the gross revenue less the technical cost of production, which can also be considered as the profit, so the summary lines represent profit share. Note that in some instances, the contractor's fee may be liable to tax.

A variation on the pure service contract with a fixed fee is to base the contractor fee on the technical costs spent by the contractor. In such a system, applied in instances in Iran and Iraq, and known as a 'buy-back' scheme, the contractor is reimbursed for the technical costs and also rewarded a fixed percentage of the investment per annum. Table 6.11 illustrates this with the reward being 50% of the annual investment in technical costs.

Table 6.11 Outline of a buy-back agreement.

	Oil price $20/bbl	Oil price $800/bbl
Gross revenue ($/bbl)	20.0	80.0
Technical costs ($/bbl); reimbursed	−5.0	−5.0
Net revenue ($/bbl)	15.0	75.0
Contractor fee ($/bbl)	−2.5	−2.5
Government take ($/bbl)	12.5	72.5
Contractor share of net revenue	=2.5/15.0 = 16.7%	=2.5/75.0 = 3.3%
Government share of net revenue	=12.5/15.0 = 83.3%	=72.5/75.0 = 96.7%

The buy-back scheme may have an unintended consequence. Since the contractor is reimbursed for the costs and additionally rewarded by an uplift on those same costs, there may be less incentive to control expenditure. In the example above, the uplift is 50% of the costs, which appears an attractive return on the investment, and therefore tempting the contractor to escalate costs — a tendency which would need to be monitored by the government. The buy-back term refers to the fact that over the period of the contract the government has repaid all costs and thus bought back the assets, ultimately becoming the owner of those assets, much as at the end of a PSC.

6.4.4.2 Risk service contracts

In a pure service contract, the contractor is not taking any risk with regard to oil price as the fee ($/bbl) relates to production only.

In a risk service contract the contractor's fee is related not to production but to gross revenues and hence will vary with oil price as illustrated in Table 6.12 in which the contractor's fee is assumed to be 30% of gross revenue. Technical costs are not reimbursed in this case, so the assumption is that the contractor's fee is sufficient to cover its investment at the prevailing oil price, hence the term risk in this form of service contract.

In this example, at $20/bbl, if the contractor's technical costs were $6.0/bbl, there would be zero net cash flow, thus emphasising the risk to the contractor in a risk service contract. Similarly, if the oil price were to drop to $16.7/bbl with a $5/bbl technical cost, the contractor's fee would be $16.7 × 30% = $5/bbl, again leaving zero net cash flow to the

Table 6.12 Contractor and government take under a risk service contract at different oil prices.

	Oil price $20/bbl	Oil price $80/bbl
Gross revenue ($/bbl)	20.0	80.0
Technical costs ($/bbl); not reimbursed	−5.0	−5.0
Net revenue ($/bbl)	15.0	75.0
Contractor fee at 30% of gross revenue ($/bbl)	−6.0	−24.0
Contractor net cash flow ($/bbl)	1.0	19.0
Government take ($/bbl)	14.0	56.0
Contractor share of net revenue	=1.0/15.0 = 6.7%	=19.0/75.0 = 25.3%
Government share of net revenue	=14.0/15.0 = 93.3%	=56.0/75.0 = 74.7%

contractor. On the other hand, the risk service contract favours the contractor in the case of high oil price. From the government's perspective, a risk service contract is a regressive form of taxation since the government share of net revenue decreases at higher oil prices or higher net revenue. The government is essentially guaranteed a 70% share of the gross revenue (in this example), much like a royalty, and it was noted in Section 6.4.2 that royalty is a regressive form of taxation.

In these examples, the net revenue has been used to illustrate the split between contractor and government. The contractor's share of the net revenue (or profit) is known as the contractor take. The government take is (1 - contractor take).

A term that is often used is the 'economic rent' of a project. In the oil and gas industry, the economic rent is received by the owner of the resource (the government or state) and equates to the gross revenue, less the cost of production, less contractor's share of the net revenue.

Economic rent is thus synonymous with the government share of net revenue in the above examples. If the contractor's share of the net revenue is considered as being reasonable, then the economic rent can be considered as the excess profit generated by the project. It is because oil and gas projects generate excess profits compared to other industries that governments have introduced creative mechanisms for capturing what they consider to be a fair economic rent, while sufficiently remunerating the contractor.

6.4.5 Association contracts

In an association contract, the IOC or private company forms what is essentially a joint venture with the NOC in the E&A activities and potential further development of an oil or gas field. An example of the association contract approach is that practised in Colombia between 1974 and 2003 when the government passed exclusive rights to ECOPETROL for exploration and development of hydrocarbons within the country and the freedom to work in association with national or foreign investors. During this period, ECOPETROL set up various association contracts using the principles of concession agreements, PSAs and risk-sharing agreements, while taking a share of the investment in each case, effectively a joint venture. In 2003, the government set up the National Hydrocarbon Agency (ANH) to take over the role from ECOPETROL and establish a new set of agreements.

It may be the case that when the government participates in the contract, the contractor carries the government's share, or part share, of the investment through certain periods of the project, such as E&A and sometimes the initial capital investment in development. A 50% government carry would imply that the contractor pays its own share plus 50% of the government's share through these investment periods, but would expect to recover this from cost oil or equivalent at a later stage, sometimes with an uplift as a form of compensation. Through this mechanism, the government limits its exposure if E&A is unsuccessful, and removes any obligation to its up-front share of development cost, making it attractive to countries with little free cash availability.

In association contracts, the contractor is normally the operator of the asset. This takes advantage of the technical and managerial experience the IOC brings to the venture, and allows a newly formed NOC to learn by shadowing activities.

6.4.6 Joint ventures

Joint ventures (JVs) are not a form of fiscal arrangement, but rather a collaboration between different parties in the investment. This can take the form of IOCs, such as Shell and Esso forming a JV partnership for development of oil and gas fields in the North Sea, or a collaboration between IOCs and NOCs or government representatives, such as that between Shell and the Omani government who formed a JV company called Petroleum Development Oman (PDO). The NOC in a JV operation is simply a partner in that investment, paying its share of investment and taking its share of the contractor's take, according to its equity percentage. Recall that each party's entitlement is the contractor's take (as defined by the fiscal terms) multiplied by its equity share (or working interest).

6.4.7 Comparison of fiscal systems

Table 6.13 compares some of the features of the fiscal systems described.

Table 6.13 Comparison of main petroleum fiscal systems.

	Concessions	Production Sharing Contract (PSC)	Service Agreement (SA)
Basis	Tax and royalty	Cost oil recovery and profit share	Fixed fee to IOC
Title to hydrocarbons	IOC[a]	NOC[b]	NOC
Relative global frequency	44%	48%	8%
Funding of project	IOC 100%	IOC 100%	IOC 100%
Government participation	Rare	Yes	Yes
Government control	Low	High	High
IOC control	High	Moderate	Low
Ownership of equipment	IOC	NOC	NOC
IOC ownership of hydrocarbons	Total production minus royalty	Cost oil plus IOC share of profit oil	None
IOC typical total share of production	ca. 85%–90%	ca. 50%–60%	None

[a]IOC — International oil company known as contractor in PSC or PSA.
[b]NOC — National oil company representing host government.

Fig. 6.15 indicates the distribution of fiscal systems around the world, but note that this may vary over time. It is apparent that the traditional regions of the world producing hydrocarbons favoured concessionary terms, whereas regions developing the resources more recently have favoured PSC terms. Hybrid systems also appear, such as having royalty as part of a PSA, or some geological horizons in the same region being governed by concession terms while others by risk service contracts. Useful references for keeping up to date with the detail of the fiscal terms around the world are online publications by EY and Deloitte (Refs. [19,20]).

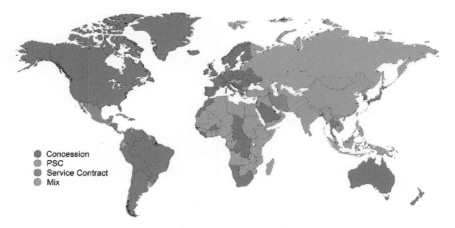

Figure 6.15 Fiscal systems around the world.

6.5 Constructing the project cash flow

The project cash flow is the basis for the economic evaluation of any project, from which we can derive a set of indicators or metrics to determine the attractiveness of the project. This section will illustrate the construction of a project cash flow under a concessionary (tax and royalty) agreement and under PSC terms. In the initial examples, the effects of inflation and the process of discounting to incorporate the time value of money will be set aside, to be addressed in Sections 6.6 and 6.7.

6.5.1 Project cash flow under tax and royalty terms

Table 6.14 shows a project cash flow under simple tax and royalty terms. Commercial and fiscal assumptions, along with opex inputs are shown in green areas, and the cash flow items are highlighted in yellow. The technical and commercial inputs have been purposely kept to round figures for clarity, but are broadly indicative of typical tax and royalty terms. The production profile is input in millions of barrels per year, but is often input as Mb/d and then converted to MMb/year (multiplying by 365.25 days per year). Care is required to ensure that the production profile includes the appropriate plant uptime and yield of sales oil, which may differ from the technical profile generated from the reservoir engineer's modelling, as referenced in Section 6.2.

Table 6.14 Project cash flow under tax and royalty terms.

Commercial and fiscal assumptions	Tax and Royalty terms		
Oil price	50	$/bbl	
Inflation	0	% p.a.	
Fixed opex	5	% cumulative capex	
Variable opex	5	$/bbl	
Royalty rate	10	%	
Tax rate	60	%	
Capital allowance	25	% p.a. straight line	

YEAR				1	2	3	4	5	6	7	8	9	TOTAL
Oil Price	50	$/bbl											
ANNUAL PRODUCTION		MMbbl	0	0	10	20	20	15	10	5	1	80
GROSS REVENUE		$m			500	1000	1000	750	500	250	50	4000
PLATFORM CAPEX		$m	160	250								410
FACILITIES CAPEX		$m		200	150							350
DRILLING CAPEX		$m	40	50	50	40						180
TOTAL CAPEX		$m	200	500	200	40						940
Cumulative Capex			200	700	900	940	940	940	940	940	940	
Capital allowance - Year 1 capex		$m			50	50	50	50				200
Capital allowance - Year 2 capex		$m			125	125	125	125				500
Capital allowance - Year 3 capex		$m			50	50	50	50				200
Capital allowance - Year 4 capex		$m				10	10	10	10			40
TOTAL CAPITAL ALLOWANCE		$m			225	235	235	235	10			940
Unit Variable opex	5	$/bbl										
VARIABLE OPEX		$m			50	100	100	75	50	25	5	400
FIXED OPEX		$m			45	47	47	47	47	47	47	280
TOTAL OPEX		$m			95	147	147	122	97	72	52	680
Royalty Rate	10	%										
ROYALTY		$m	0	0	50	100	100	75	50	25	5	400
TAX DEDUCTIONS		$m	0	0	370	482	482	432	157	97	57	2020
TAXABLE INCOME		$m	0	0	130	518	518	318	343	153	-7	1980
Taxrate	60	%										
TAX		$m	0	0	78	310.8	310.8	190.8	205.8	91.8	-4.2	1188
GROSS REVENUE		$m	0	0	500	1000	1000	750	500	250	50	4000
CAPEX + OPEX (TECHNICAL COST)		$m	200	500	295	187	147	122	97	72	52	1620
GOVERNMENT TAKE		$m	0	0	128	410.8	410.8	265.8	255.8	116.8	0.8	1588
NET CASH FLOW		$m	-200	-500	77	402.2	442.2	362.2	147.2	61.2	-2.8	792
CUMULATIVE NET CASH FLOW		$m	-200	-700	-623	-220.8	221.4	583.6	730.8	792	789.2	

The logic of the cash flow calculation will follow, starting from the commercial and technical assumptions at the top of Table 6.14.

Oil price is assumed to be flat MOD, meaning no escalation of oil price (see Fig. 6.5). Inflation and hence cost escalation are assumed to be zero in this example.

Total opex is represented by fixed plus variable elements. Annual **fixed opex** is 5% of the cumulative capex, so note that this becomes flat once all capex has been invested, but does not disappear. Throughout field life the operator must continue to maintain and run the large plant that has been constructed.

Variable opex is modelled as $5/bbl of oil production, and thus decreases as the production declines. This is a common assumption in the cash flow model, but could be challenged — will operating costs really decline as the plant gets older and requires more maintenance, and wells start to deliver unwanted products such as water, sand and possibly corrosive fluids such as H_2S? For a water-flooded oil field, it may be more logical to base the variable opex on gross liquid production rather than oil production, as this will remain flat or possibly increase over the producing life. Nevertheless, the assumption of declining variable opex is common.

The **tax rate** represents a single tax at 60%, to be applied to the taxable income. In many cases, there may be several layers of taxation, with the first tax becoming an allowance against the second. For example, when applied in the United Kingdom, PRT was charged on PRT taxable income, but then became an allowance against CT. So when the PRT rate was 50% and CT was 30%, then

$$\text{effective total tax rate} = 1 - [(1 - \text{PRT}) \times (1 - \text{CT})]$$
$$= 1 - [(0.5) \times (0.7)] = 0.65 = 65\%$$

In fact, after 2015 PRT reduced to 0% on all fields in the United Kingdom, as the fiscal system was relaxed in response to declining oil price.

Capital allowance in this example is assumed to follow the straight line method (the simplest of the methods) at 25% per annum, so claimed over 4 years from the first date of each year's expenditure on this project. The sum of the capital allowance equals the total capex, so the allowances fully recover the capex investment over the project lifetime. The example assumes no cross-field allowances, so this project is effectively ring-fenced, or else the investor is a newcomer to the region. Note that the capital allowance method and all fiscal terms are dictated by the host government, and not at the discretion of the operator.

Moving on to the cash flow calculation, performed annually, and presented in the table in millions:

$$\text{gross revenue}(\$) = \text{production}(\text{bbl}) \times \text{oil price}(\$/\text{bbl})$$

The **total capex** is composed of platform, facilities and drilling costs and is aligned in time with the project schedule, but nevertheless capex is mostly an up-front project cost. The cumulative capex is recorded as it is required for the later calculation of the fixed opex. The capital allowance calculation is based on 25% per annum straight line, starting in the year of first revenue. Note that capex is a cash flow item (money is physically paid), whereas capital allowance is part of the calculation required for tax calculations, but is not a cash flow item.

Total opex is composed of the fixed and variable elements, noting in this example that the lifetime opex is similar to the total capex. This is common in many projects, but can be overlooked if focussing on up-front costs and initial project exposure only. Opex is a cash flow item. Often the sum of capex and opex is referred to as the technical cost, or individually as development and lifting cost. In general, on an annual basis

$$\text{total opex}(\$) = A(\%) \times \text{cumulative capex}(\$)$$
$$+ B(\$/\text{bbl}) \times \text{production}(\text{bbl})$$

where factors A and B depend upon the geography and the technical complexity of the project. These factors can be estimated using costing tools such as QUE$TOR or based on the operator's regional knowledge.

Royalty is paid to the host government, and is a cash flow item, based on a percentage of the gross revenue. No costs are allowable before paying royalty, and it is often referred to as being taken 'off the top of the barrel'. It is a regressive form of taxation, meaning that as costs of production increase, royalty becomes more punitive for the operator.

$$\text{royalty}(\$) = \text{gross revenue}(\$) \times \text{royalty rate}(\%)$$

Tax deductions, also known as fiscal costs or fiscal allowances, are offsets against gross revenues in the calculation of **taxable income**. These deductions are allowances for the cost of production and include royalty paid to the government plus opex, plus the capital allowance. Note that opex can be claimed as a tax deduction in the year of expenditure, whereas capex is claimed as a tax deduction over a period of time through the capital allowance. Generally, the operator prefers to claim the allowance as early as possible, but the fiscal terms will dictate what expenditure is classed as opex and capex. A tangible item with a useful lifetime of more than a year is typically classed as capex. However, fiscal incentives for investment may be offered, for example, by allowing a well cost to be claimed in the year of expenditure, effectively treating it as opex in the calculation of tax deductions. This is not the case in the example above. The tax deduction is not a cash flow item, but a necessary step in calculating taxable income. On an annual basis

tax deductions($) = royalty($) + opex($) + capital allowance($)

Taxable income is the basis for the tax calculation, so that tax is charged on taxable income at the total tax rate. Again, taxable income is not a cash flow item, whereas tax is a cash flow item, paid physically to the host government. In the example, only one overall tax is applied.

taxable income($) = gross revenue($) − tax deductions($)
tax($) = taxable income($) × tax rate(%)

Note that in Year 9 the taxable income has become negative. This is because the gross revenue is less than the sum of the royalty plus opex, and the pre-tax cash flow is now negative. The **economic lifetime** of the project has now become apparent, and the project should stop, thus incurring decommissioning costs. The final year of the cash flow has been greyed out in the example, and cumulative figures do not include any contribution from Year 9, assuming that production stops at the end of Year 8. Decommissioning costs will be addressed in Chapter 7.

The final section of the example in Table 6.14 reports the project cash flow items, starting with a repeat of the gross revenue, and then deducting the technical cost (capex + opex) and the government take (royalty + tax) to yield the net cash flow, which can also be expressed as the cumulative net cash flow.

technical cost($) = capex + opex($)
government take($) = royalty + tax($)
net cash flow($) = gross revenue − technical cost
− government take($)

The share of the gross revenue is now apparent. Of the $4000 m of gross revenue, $1620 m (40.5%) was consumed by the technical costs, the government has received $1588 m (39.7%) from royalty plus tax and the operator has received a net cash flow of $792 m (19.8%). The government's share can be considered as the economic rent — refer to Section 6.4.4 for the oil field definition of economic rent.

In terms of the net revenue (gross revenues less technical costs, $4000 - 1620 = 2380$ m) the government has received 1588 m (66.7%) and the operator $792 m (33.3%).

Fig. 6.16 shows the components of the net cash flow.

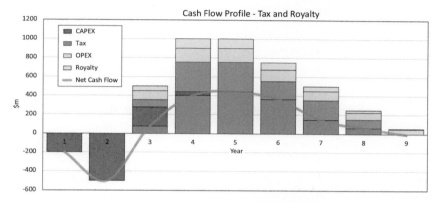

Figure 6.16 Components of the project annual cash flow under tax and royalty terms.

This cash flow is expressed on a cumulative basis in Fig. 6.17.

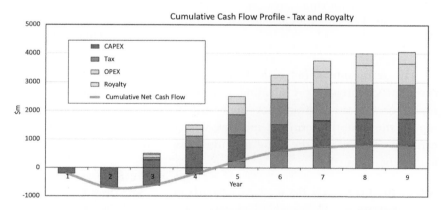

Figure 6.17 Project cumulative cash flow under tax and royalty terms.

Some simple project economic indicators from the annual and cumulative cash flow analysis are shown in Table 6.15. These indicators are discussed further in Chapter 7.

Table 6.15 Economic indicators from the net cash flow.

Project economic indicator	Calculated from	Interpretation
From the annual cash flow:		
Economic lifetime (year)	First-year net cash flow becomes permanently negative	Project should stop here and prepare for decommissioning
Pay-as-you-go time (year)	First year in which net cash flow becomes positive	Project no longer requires investment — it has become self-funding
From the cumulative cash flow:		
Maximum exposure ($)	Lowest value on the cumulative net cash flow	Level of cash required to carry out project
Payback time (year)	Time when cumulative net cash flow returns to zero after the initial investment	Initial investment just recovered at this point in time
Ultimate cash surplus ($)	Final cumulative net cash flow at the economic lifetime	Undiscounted return from the project, excluding decommissioning cost

The term annual net cash flow relates specifically to the petroleum economics approach to evaluating a project. However, when accountants report or forecast performance on behalf of the company in an annual report, they use the term net income, which may also be called profit or earnings. This is not the same as annual net cash flow, and Section 7.6 makes the distinction.

6.5.2 Project cash flow under Production Sharing Contract terms

Table 6.16 shows the same technical project as described under the simple tax and royalty fiscal system, but now using typical PSC terms. Commercial and fiscal assumptions, along with opex inputs, are shown in green areas, and the cash flow items are highlighted in yellow. The technical and commercial inputs are purposely kept to round figures for clarity, but their values are indicative of typical PSC terms.

Project cash flow

Table 6.16 Cash flow calculation using PSC terms.

Commercial and fiscal assumptions PSC terms		
Oil price	50	$/bbl
Inflation	0	% p.a.
Fixed opex	5	% cumulative capex
Variable opex	5	$/bbl
Royalty rate	10	%
Cost oil allowance	50	% of gross revenue
Profit oil split to contractor	40	%
Tax rate on profit oil	40	%

YEAR			1	2	3	4	5	6	7	8	9	TOTAL
Oil Price	50	$/bbl										
ANNUAL PRODUCTION		MMbbl	0	0	10	20	20	15	10	5	1	80
GROSS REVENUE		$m			500	1000	1000	750	500	250	50	4000
PLATFORM CAPEX		$m	160	250								410
FACILITIES CAPEX		$m		200	150							350
DRILLING CAPEX		$m	40	50	50	40						180
TOTAL CAPEX		$m	200	500	200	40						940
Cumulative Capex			200	700	900	940	940	940	940	940	940	
Unit Variable opex	5	$/bbl										
VARIABLE OPEX		$m			50	100	100	75	50	25	5	400
FIXED OPEX		$m			45	47	47	47	47	47	47	280
TOTAL OPEX		$m			95	147	147	122	97	72	52	680
Royalty Rate	10	%										
ROYALTY		$m	0	0	50	100	100	75	50	25	5	400
Cost oil maximum (COM)		$m	0	0	250	500	500	375	250	125	25	
COM - Opex		$m	0	0	155	353	353	253	153	53	−27	
Capex recovery		$m	0	0	155	353	353	79	0	0	−27	940
Unrecovered capex		$m	200	700	745	432	79	0	0	0	27	
TOTAL COST OIL RECOVERY		$m			250	500	500	201	97	72	25	1620
Profit oil split to contractor	40	%										
Profit oil		$m	0	0	200	400	400	474	353	153	20	1980
Profit oil to government		$m	0	0	120	240	240	284.4	211.8	91.8	12	1188
PROFIT OIL TO CONTRACTOR		$m	0	0	80	160	160	189.6	141.2	61.2	8	792
Tax rate on profit oil	40	%										
TAX		$m	0	0	32	64	64	75.84	56.48	24.48	3.2	316.8
GROSS REVENUE		$m	0	0	500	1000	1000	750	500	250	50	4000
CAPEX + OPEX (TECHNICAL COST)		$m	200	500	295	187	147	122	97	72	52	1620
TOTAL COST OIL RECOVERY		$m	0	0	250	500	500	201	97	72	25	1620
PROFIT OIL TO CONTRACTOR		$m	0	0	80	160	160	189.6	141.2	61.2	8	792
TAX		$m	0	0	32	64	64	75.8	56.5	24.5	3.2	316.8
NET PROFIT OIL TO CONTRACTOR		$m	0	0	48	96	96	113.8	84.7	36.7	4.8	475.2
CONTRACTOR NET CASH FLOW		$m	−200	−500	3	409	449	192.8	84.7	36.7	−22.2	475.2
CUMULATIVE NET CASH FLOW		$m	−200	−700	−697	−288	161	353.8	438.5	475.2	453.0	

The assumptions and calculations in the PSC example follow that of the tax and royalty example as far as the royalty calculation. Royalty is again assumed to be 10% of gross revenue. Thereafter the calculation differs and is explained systematically as follows.

Cost oil is made available for the contractor to recover opex and capex and is allocated as a maximum of 50% of the value of annual production. Hence the **COM** is 50% of gross revenue, on an annual basis. The contractor can claim up to the annual COM, but not more.

$$\text{cost oil maximum}(\text{COM})(\$) = \text{gross revenue}(\$)$$
$$\times \text{ cost oil allowance}(\%)$$

In the first 2 years, there is no cost oil available, as production starts only in Year 3. Once cost oil becomes available, firstly the opex is claimed from COM, leaving the balance available for capex recovery

$$\text{cost oil available for capex recovery}(\$) = \text{COM}(\$) - \text{opex}(\$)$$

The **capex recovery** is then claimed from the cost oil available for capex recovery, but in the early years of production, there is insufficient cost oil available to recover all of the capex spent to date. It is therefore necessary to track the **unrecovered capex** and roll this forward until all capex has been recovered. So for each year

$$\text{unrecovered capex}(\$) = \text{capex spent this year}(\$)$$
$$+ \text{ unrecovered capex from previous year}(\$)$$
$$- \text{ capex recovery this year}(\$)$$

In this example, all the capex is recovered from cost oil by the end of Year 6. The total cost oil recovery is the sum of the opex recovery plus the capex recovery. In later years this sum is less than the COM, since there is then ample COM to recover the opex. This leaves more profit oil to be split between the contractor and the government.

$$\text{total cost oil recovery}(\$) = \text{opex}(\$) + \text{capex recovery}(\$)$$

The example assumes that there is always sufficient cost oil available to recover the annual opex, and this is usually the case. However, if the annual opex were to exceed the COM, then the unrecovered opex would need to be carried forward in the same way as capex in the example.

After deduction of royalty and distribution of cost oil, the remainder of the gross revenue is profit oil, which is then shared between the government and the contractor, the latter taking the 40% split of profit oil in the example.

$$\text{profit oil}(\$) = \text{gross revenue} - \text{royalty} - \text{cost oil used}(\$)$$
$$\text{profit oil to contractor}(\$) = \text{profit oil}(\$)$$
$$\times \text{ profit oil split to contractor}(\%)$$
$$\text{profit oil to government}(\$) = \text{profit oil}(\$)$$
$$\times \text{ profit oil split to government}(\%)$$

The contractor pays tax on its share of the profit oil, at the prevailing tax rate.

tax($) = profit oil to contractor($) × tax rate on profit oil(%)

Once again, the economic lifetime of the project has been reached at the end of Year 8, since the cost oil available is insufficient to cover the opex, and the project should stop thus incurring decommissioning costs. The final year of the cash flow has been greyed out in the example, and cumulative figures do not include any contribution from Year 9, assuming that production stops at the end of Year 8. The treatment of decommissioning costs is addressed in Chapter 7.

The final section of the example in Table 6.16 reports the project cash flow items, being a repeat of the gross revenue, the technical costs (capex plus opex), the total cost oil recovery, the profit oil to the contractor and the tax on the contractor's profit oil. The net profit oil to the contractor accounts for the tax paid to the government.

net profit oil to contractor($) = profit oil to contractor($) − tax($)

Finally, the contractor annual net cash flow can be calculated, and also expressed as a cumulative net cash flow.

contractor net cash flow($) = net profit oil to contractor($)
$$+ \text{total cost oil recovery}(\$)$$
$$- \text{technical cost}(\$)$$

Of the $4000 m of gross revenue, $1620 (40.5%) was consumed by technical costs, the government has received royalty plus its share of profit oil plus tax ($400 m + $1188 m + $316.8 m = $1904.8 m) representing 47.6% of gross revenue, and the contractor's take is $475.2 m (11.9%).

In terms of the net revenue (gross revenues less technical costs, $4000 m−$1620 m = $2380 m) the government has received $1904.8 m (80.0%) and the operator $475.2 m (20.0%).

Figs. 6.18 and 6.19 show the annual and cumulative cash flow of the project.

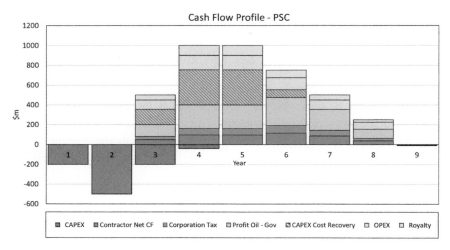

Figure 6.18 Components of the project annual cash flow under PSC terms.

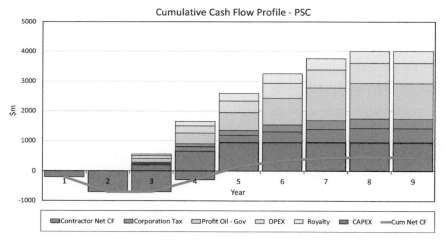

Figure 6.19 Project cumulative cash flow under PSC terms.

In these examples, the production profile, technical cost and oil price assumptions are consistent, and typical tax and royalty and PSC terms have been assumed. The net cash flow to the contractor in the two cases is not necessarily comparable. If PSC terms offered are so stringent as to make the contractor's investment unattractive, then a negotiation is required to come to agreed terms. Only when both contractor and government can accept the PSC terms will an agreement be reached and a contract signed.

Nevertheless, the simple economic indicators introduced are summarised for these two cases, as shown in Table 6.17. While the contractor's share of gross revenue and net revenue (gross revenue less technical cost) is not an economic indicator, it is included for interest.

Table 6.17 Summary of simple economic indicators from tax and royalty and PSC examples.

Project economic indicator	Tax and Royalty	PSC
From the annual cash flow:		
Economic lifetime (year)	8	8
Pay-as-you-go time (year)	3	3
From the cumulative cash flow:		
Maximum exposure ($m)	−700	−700
Payback time (year)	5	5
Contractor's ultimate cash surplus ($m)	792	475
For interest (but not comparison):		
Contractor share of gross revenue (%)	19.8	11.9
Contractor share of net revenue (%)	33.3	20.0

6.6 Discounting − the time value of money

The cash flow analysis of the examples in this chapter so far has considered that the value of the cash in each year is the same. The cumulative net cash flow has assumed that the dollar spent in Year 1 is equivalent to the dollar earned in Year 8. This is a simplistic view, as it does not account for what is commonly termed the 'time value of money' − although this is not a particularly self-explanatory phrase.

Offered the choice between $100 in hand today or the promise of $100 in 8 years' time, you would opt for the money today. It may be because of the certainty of the money today versus the risk of not receiving the promised sum in 8 years' time, but even if the future sum were a guarantee, you would take the money today. Why is this?

A common argument uses the effect of inflation, expected to increase the cost of goods over time, and therefore reduce the purchasing power of the $100 in Year 8. If a loaf of bread costs $1 today, and assuming an annual inflation rate of 3%, the price of a loaf of bread in Year 8 would increase to $\$1 \times (1 + 0.03)^8 = \$1.27$. The $100 in Year 8 would purchase only 79 loaves ($100/$1.27 per loaf), whereas one could buy 100 loaves with the $100 today. This seems a logical explanation to preferring the $100 today rather than $100 in 8 years' time. However, imagine that there is no inflation − which would you prefer? Still the $100 today.

The effect of inflation is not the real driver in the decision, but rather the opportunity to invest the $100 today and watch it grow. If one could invest the $100 at a compound interest rate of say 8%, then the value of the $100 today would grow as shown in Table 6.18.

Table 6.18 Effect of compound interest on an investment.

Time (year)	Initial investment ($)	Compound interest calculation 8% p.a.	Year-end value ($)	Name of monetary sum
0	100			Present value
1		$= 100 \times (1 + 0.08)^1$	108.0	
2		$= 100 \times (1 + 0.08)^2$	116.6	
3		$= 100 \times (1 + 0.08)^3$	126.0	
4		$= 100 \times (1 + 0.08)^4$	136.0	
5		$= 100 \times (1 + 0.08)^5$	146.9	
6		$= 100 \times (1 + 0.08)^6$	158.7	
7		$= 100 \times (1 + 0.08)^7$	171.4	
8		$= 100 \times (1 + 0.08)^8$	185.1	Future sum

Compound interest is surprisingly powerful. Albert Einstein is quoted as saying that 'Compound interest is the eighth wonder of the world. He who understands it, earns it; he who doesn't, pays it'. It may be surprising that the value of an initial investment made at a compound interest of 10% per annum will double in just over 7 years $(1 + 0.1)^7 = 1.95$.

The starting point for the example in Table 6.18 is the beginning of Year 1 (time $= 0$), being the **reference date**. The year-end values occur at the end of each year, so the investment of $100 at the start of Year 1 accrues to a future sum of $108 at the end of Year 1 and to $185.1 by the end of Year 8.

Assuming there were no other options for investment, you should be ambivalent about the choice of receiving the $100 today, which you can invest at 8% annual compound interest rate, or receiving $185.1 in 8 years' time. In other words, the future sum of $185.1 at the end of Year 8 is equivalent to the present value (PV) of $100 today, assuming the limitation on investment options. In the example, the investment at the reference date escalates with compound interest, and the calculation is simple, as shown in Table 6.18.

To reconcile the value of all future annual net cash flows from the project, the opposite of the compound interest formula is applied to each future net cash flow to calculate its **PV**. This process is called **discounting**. So, the PV of a future sum of money (C_t) can be determined using the following formula.

$$\text{Present Value(PV)} = \frac{\text{future sum}(C_t)}{(1+r)^t} (\$)$$

where
  $r =$ the discount rate (fraction)
  $t =$ time between the future sum occurring and the reference date (years)

Using the above example, the PV of $185.1 occurring in Year 8 is

$$PV = \frac{185.1}{(1+0.08)^8} = 100(\$)$$

This formula is applied to every future net cash flow in the project. The PVs of the early years' cash flows will be negative, the later years' PVs will be positive. The sum of all PVs is the net present value (NPV) of the project.

For the tax and royalty example, using a discount rate of 8%, the calculation of project NPV is shown in Table 6.19 and is noted as NPV(8) or NPV_8 to indicate that an 8% discount rate has been used. The factor $1/(1+r)^t$ is known as the **discount factor**. This is of course independent of the cash flow, and when multiplied by the cash flow yields the PV.

$$\text{Present Value(PV)} = \frac{1}{(1+r)^t} \times \text{future sum}$$

The discount factor for each year is included in Table 6.19 for clarity. The net cash flow in Year 9 has been excluded from the NPV calculation since the economic lifetime was established as being the end of Year 8. Decommissioning cost will be included in Chapter 7.

Table 6.19 Calculation of net present value (NPV).

YEAR		1	2	3	4	5	6	7	8	9	TOTAL
NET CASH FLOW	$m	−200.0	−500.0	77.0	402.2	442.2	362.2	147.2	61.2	−2.8	792.0
Discount rate	8%										
Full-year discount factor		0.926	0.857	0.794	0.735	0.681	0.630	0.583	0.540		
PV	$m	−185.2	−428.7	61.1	295.6	301.0	228.2	85.9	33.1	391.1	NPV(8)

	Excel formula	=NPV(8%,D3:K3)	391.1

There is a simple formula in Excel that will calculate NPV, and requires the discount rate and the cells containing the net cash flow figures, in this case cells D3:K3 as shown in Table 6.19. Care is required to enter the discount rate as percentage (8%) or fraction (0.08), as input of 8 will discount at 800%. Excel assumes that the cash flows occur at the end of each period, an assumption known as full-year discounting.

It is, however, an unreasonable assumption that the cash flows occur at the end of each year, or indeed at the beginning of each year. In fact, cash settlements are likely to occur at the end of each month, and so in theory we should be discounting on a monthly basis. This is, however, impractical, and as a compromise, a common practice is to assume the annual net cash flow occurs in the middle of each year, and to discount back from this time to the reference date, known as mid-year discounting (Fig. 6.20).

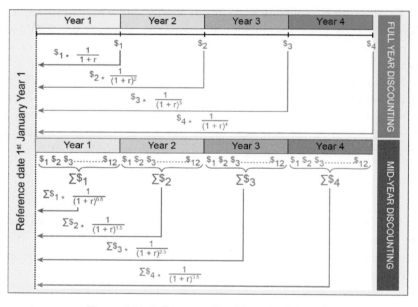

Figure 6.20 Full-year and mid-year discounting.

To perform mid-year discounting the discount factor should be adjusted to express each time period as (t − 0.5) as shown in the following formula.

$$\text{Present Value(PV)} = \frac{1}{(1+r)^{(t-0.5)}} \times \text{future sum}$$

Excel can be adjusted to do this by adapting the NPV equation to = NPV(8%,D2:K3)*(1 + 0.08)^0.5.

This essentially moves the complete full-year discounted NPV one half year earlier, and uplifts the value, as shown in Table 6.20. Many companies adopt mid-year discounting as being a more accurate representation of the project cash flows, but it is still an approximation. Consistency of method across projects within the company is important.

Table 6.20 Calculation of net present value (NPV) using mid-year discounting.

YEAR		1	2	3	4	5	6	7	8	9	TOTAL
NET CASH FLOW	$m	-200.0	-500.0	77.0	402.2	442.2	362.2	147.2	61.2	-2.8	792.0
Discount rate	8%										
Mid-year discount factor		0.962	0.891	0.825	0.764	0.707	0.655	0.606	0.561		
PV	$m	-192.5	-445.5	63.5	307.2	312.8	237.2	89.3	34.4	406.4	NPV(8)

Excel formula	=NPV(8%,D3:K3)*(1+0.08)^0.5	406.4

6.6.1 Cost of capital

The discount rate used in the NPV calculation is the cost of capital to the investor, assumed to be 8% in the example.

The cost of capital represents the average cost of borrowing the funds to invest in the project. For a listed company, funds will be provided as a mixture of loans from banks, called debt, and money invested by the shareholders when they purchase shares, called equity. The cost of borrowing from the bank is the interest payable on the loan, which is tax deductible in calculating accounting profit. The cost of borrowing from shareholders is the dividend paid on the shares.

The Weighted Average Cost of Capital (WACC) method uses the following formula to estimate the cost of capital.

$$\text{WACC} = \frac{D \times K_D \times (1 - \text{corporation taxrate}) + E \times K_E}{D + E}$$

where

  WACC = weighted average cost of capital (%)
  D = market value of debt ($)
  K_D = cost of debt (%)
  E = market value of equity ($)
  K_E = cost of equity (%)

Applying the WACC method can be complicated, since there are a number of types of bank loan (long-term and short-term bonds) each with their own rates depending on the pecking order of payment and associated level of risk. The cost of debt will be determined by each lender based on a risk-free rate (say a published government treasury bond rate of 3%) plus an uplift (say 4%) depending on the credit rating of the company and the ratio of debt to equity. The interest payments are tax deductible (say 30% corporate tax rate), so that

$$\text{cost of debt} = (3\% + 4\%) \times (1 - 30\%) = 4.9\%$$

The same is true of the equity as the different types of share (ordinary, preference) also carry different costs. The capital asset pricing model (CAPM) can be used to estimate the cost of equity

$$\text{cost of equity} = \text{risk free rate of return} + \text{beta} \\ \times (\text{market rate of return} - \text{risk free rate of return})(\%)$$

In the CAPM the risk-free rate of return is what a shareholder might expect by investing in a risk-free vehicle such as a government bond, say 3%. The market rate of return is what the shareholder demands for investing in the particular company (say 7%). This will be higher than the risk-free rate as the shareholder is low in the pecking order for payout if the company struggles financially. The beta value is a measure of the variation of the company shares compared to the market sector, based on a regression of the company's share price variation using historical data. A beta greater than 1.0 means the share price is volatile and reacts more strongly than the market sector, and a beta less than 1.0 means the share price reacts less than the market sector, a so-called defensive asset. If the beta for a company is 1.5, then

$$\text{cost of equity} = 3\% + 1.5 \times (7\% - 3\%) = 9\%$$

If the debt is listed, then the fair value of the debt can be used as a market value of the debt, otherwise the balance sheet value (book value) of all debt is used. The market value of the equity is the number of outstanding shares multiplied by the share price. This can be complicated when there are different types of shares issued. Ref. [24] is useful for further reading.

The WACC approach is detailed and requires a financial understanding, but an approximation can be used by a book value approach, taking data from the published accounts.

$$\text{cost of capital}(\%) = \frac{\text{interest on debt}(\$) \times (1 - \text{corporation tax rate}) + \text{shareholder dividend}(\$)}{\text{market value of debt}(\$) + \text{market value of share capital}(\$)}$$

The annual interest and dividend payments can be read from the profit and loss account, along with the average interest rate paid on profit before tax. The total debt can be found on the balance sheet. The market value of the share capital is the number of outstanding shares multiplied by the share price, also known as the market capitalisation.

The annual reports should be reviewed back over a number of years to estimate an average WACC, as dividend payments and share price may vary depending on the annual performance of the company.

If listed companies are borrowing on a corporate basis and then distributing investment funds across projects, then the WACC should be the same for all projects. This is in contrast to individual project finance, where capital is raised against a specific project. Large companies will generally raise capital corporately as the strength of the corporate balance sheet allows the negotiation of lower interest rates on debt. Typically, the WACC to IOCs in the upstream sector is 7%–10%. Recall that the example discount rate used in this section was 8%, deemed to be a typical cost of capital for such companies.

6.6.2 Opportunity cost

Some texts refer to the discount rate being set equal to the **opportunity cost**, rather than the cost of capital. The opportunity cost of undertaking one project is the loss of return foregone by not undertaking an alternative. Therefore, if Project A is undertaken in preference to Project B, the funds invested in Project A cannot be invested in Project B. On the assumption that Project B could provide a certain return, then this is the opportunity cost foregone by undertaking Project A.

By discounting at an opportunity cost (say 8%), it is implied that Project B could earn a return equal to the cost of capital (8%).

The two approaches to explaining discounting come together if we assume that by not investing in Project A we could invest in the alternative Project B which earns a return equal to the cost of capital. Project B may be putting the investment capital into a bank that guarantees an 8% interest rate. In other words, we are assuming that the cost of borrowing (8%) is equal to the rate of return of the alternative investment (8%).

When discounting at the cost of capital, the NPV calculated is therefore the value added by the project after repaying the cost of the investment capital.

6.6.3 Internal rate of return

In principle, a future cash flow sum can be discounted at any chosen discount rate. Table 6.21 shows the net cash flow in the cash flow example, discounted at 0%, 8%, 15%, 20%, 25%, 30%, calculated using Excel formulas for each year, and again using the NPV formula, both assuming mid-year discounting. The PV for each year is also shown for each discount rate, forming the basis for Fig. 6.21. Year 9 is excluded as being beyond the economic lifetime.

Table 6.21 NPV calculated at various discount rates.

YEAR		1	2	3	4	5	6	7	8	TOTAL	
NET CASH FLOW	$m	−200.0	−500.0	77.0	402.2	442.2	362.2	147.2	61.2	792.0	
PV(0)	$m	−200.0	−500.0	77.0	402.2	442.2	362.2	147.2	61.2	792.0	NPV(0)
PV(8)	$m	−192.5	−445.5	63.5	307.2	312.8	237.2	89.3	34.4	406.4	NPV(8)
PV(15)	$m	−186.5	−405.4	54.3	246.6	235.8	167.9	59.3	21.5	193.4	NPV(15)
PV(20)	$m	−182.6	−380.4	48.8	212.5	194.7	132.9	45.0	15.6	86.5	NPV(20)
PV(25)	$m	−178.9	−357.8	44.1	184.2	162.0	106.2	34.5	11.5	5.8	NPV(25)
PV(30)	$m	−175.4	−337.3	40.0	160.6	135.8	85.6	26.7	8.6	−55.6	NPV(30)
			Excel formula	=NPV(0%,D3:K3)*(1 + 0.00)^0.5						792.0	NPV(0)
				=NPV(8%,D3:K3)*(1 + 0.08)^0.5						406.4	NPV(8)
				=NPV(15%,D3:K3)*(1 + 0.15)^0.5						193.4	NPV(15)
				=NPV(20%,D3:K3)*(1 + 0.20)^0.5						86.5	NPV(20)
				=NPV(25%,D3:K3)*(1 + 0.25)^0.5						5.8	NPV(25)
				=NPV(30%,D3:K3)*(1 + 0.30)^0.5						−55.6	NPV(30)

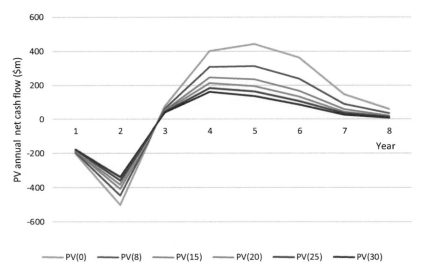

Figure 6.21 PV annual net cash flow at various discount rates

Fig. 6.21 shows the impact of discounting future sums. The further from the reference date and the higher the discount rate assumed, the more the future sum diminishes in PV.

Fig. 6.22 shows the relationship between the NPV and the discount rate — a plot called the PV profile and useful in introducing the project **internal rate of return** (IRR).

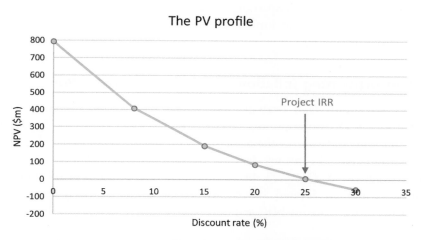

Figure 6.22 The PV profile: NPV as a function of discount rate.

By definition, the IRR of the project is the discount rate that returns an NPV equal to zero. It can be found by creating the PV profile shown in Fig. 6.22, or more simply by using the Excel Function = IRR(D3:K3) where the target range of cells contains the net cash flow. Excel uses a goal seek routine to determine the discount rate that returns a zero value for NPV.

For a cash flow as used in the example, this will be unique value. However, when more than one change of sign of the net cash flow occurs, such as in a phased project, there may be multiple solutions to IRR and this indicator must then be treated with caution.

The interpretation of IRR is important. The immediate significance is that if the IRR exceeds the cost of capital, then the project must be capable of returning a positive NPV. The project in this example has an IRR of 25.4%, which can be thought of as putting money into a bank and receiving 25.4% interest. If one could borrow at 8% cost of capital, the project would be very worthwhile, and robust to many uncertainties. The bigger the margin by which IRR beats the cost of capital, the more robust the project. Comparison of IRR with the cost of capital is paramount in its use, but further discussion of the interpretation of IRR will follow in Chapter 7.

6.7 Incorporating inflation into the project cash flow

The project cash flow example used so far has ignored the impact of inflation and decommissioning costs — both of which should be taken into account. Inflation in the western world has become accepted as the norm, and over the recent decade, low positive rates of inflation have been experienced. Fig. 6.23 shows historical inflation rates in the United Kingdom and the United States over five decades, compiled using data from the UK Office of National Statistics (ONS) and the US Bureau of

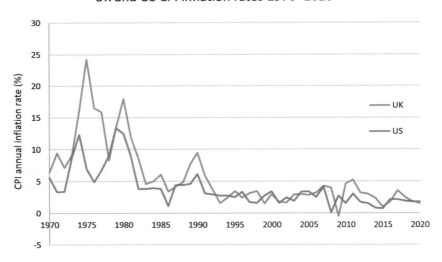

Figure 6.23 UK and US CPI inflation rates 1970—2020. *(Sources: UK ONS, US Bureau of Labor Statistics.)*

Labor Statistics. In the early period 1973—75 recession took hold in the United States followed by a period of inflationary prices, high unemployment and subsequent growth stagnation, a combination known as stagflation. Much of the western world was experiencing a similar end to the post—World War II economic expansion, and in the United Kingdom industrial relations and miners' strikes led to many companies introducing a 3-day working week to manage electric power availability. The first major oil price shock compounded the problems, and inflation peaked above 20%.

An escalating rate of inflation, increasing the price of goods and services, creates uncertainty and compels industry to reduce investment in the face of unpredictable future costs, thus slowing economic growth. To combat the situation, governments have several tools at their disposal, including monetary policy (adjusting interest rates), controlling the money supply, fiscal policy (increasing tax rates to reduce spending) and wage controls. The favoured method over the last three decades has been the application of a monetary policy under which governments task central banks with adjusting the interest rates as a control on inflation. Increasing interest rates makes investing more attractive and borrowing more expensive, leading to reduced demand and a resulting slowing of inflation. As Fig. 6.23 shows, this has been largely successful, with governors of central banks tasked to control annual inflation within a target band of around 2%—3% through this monetary approach. The target in the United Kingdom has been 2% inflation ±1%, and base interest rate has been in the hands of the Governor of the Bank of England.

The measure used in Fig. 6.23 is the Consumer Price Index (CPI), which is defined as the change in the price of a basket of goods and services that are typically purchased by specific groups of households. The basket includes food and energy, and so is linked directly to oil and gas prices.

An alternative measure recorded is the Retail Price Index (RPI), which measures the sales price of a basket of goods as it leaves the producers' premises, and is closely related to the CPI.

Inflation increases the price of goods year-on-year and forecasting the GRI has become easier to forecast as the historical rate has stabilised. Although relatively low, price escalation should be included in the capex and opex elements of the project cash flow. Fig. 6.24 introduces the steps required to incorporate inflation, and calculate what is termed the MOD net cash flow and the RT net cash flow. The steps will now be explained.

198 Petroleum Economics and Risk Analysis

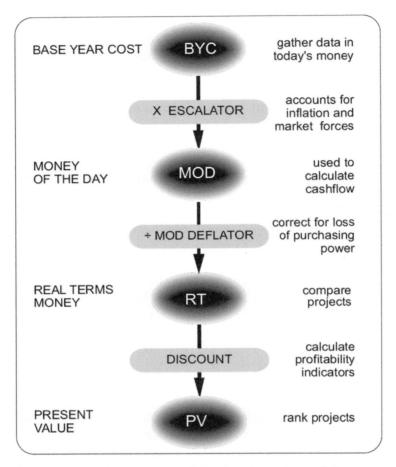

Figure 6.24 Incorporating inflation into the project cash flow.

An estimate of the future cost of an item, say a steel jacket to be built and paid for in 4 years' time, can be made by finding the cost of the jacket today from a list price and then escalating this using an assumed inflation rate. In addition to the GRI, there may be specific market factors that can be used to adjust the escalation. In the case of the jacket, the trend in price of steel or availability of fabrication yards may be of significance. The market force may be forecast to increase or decrease the future price, and the market factor will require detailed knowledge of the particular market. For example, reviewing operators' commitments to regional exploration and appraisal can inform the trend in future rates for mobile drilling rigs.

The actual cost of the jacket in 4 years' time is called the MOD, also known as nominal terms money, and is derived from the Base Year Cost (BYC) using the combined escalation factor as follows.

$$\text{MOD cost} = \text{BYC} \times (1 + \text{inflation rate})^t \times (1 + \text{market factor})^t$$

where
   MOD cost = Money of the Day cost ($); the actual future cost when item is bought
   BYC = Base Year Cost ($); what the item costs at the reference date
   inflation rate = general annual rate of inflation, based on CPI or similar (fraction)
   market factor = annual rate of cost escalation for specific item (fraction)
   t = year in which item is purchased relative to the reference date (year)

So, an item with a BYC of $200 m, an assumed inflation rate of 5% p.a. and a market factor of 2%, payable in 4 years' time, would have an MOD cost estimate of

$$\text{MOD cost}(\$) = 200 \times (1 + 0.05)^4 \times (1 + 0.02)^4 = 263(\$m)$$

For small inflation and escalation rates, an acceptable approximation is to add the market factor and GRI together.

$$\text{MOD cost}(\$) = 200 \times (1 + 0.05 + 0.02)^4 = 262(\$m)$$

Though it may be considered more accurate to escalate costs to mid-year values (as introduced in Section 6.6), this is an unnecessary complication. All cash flows will be assumed to occur at year-end and the mid-year shift will be accommodated in the final discounting step.

In addition to escalating the costs, oil price may also be escalated, as discussed in Section 6.2. Recall that assuming a flat RT oil price implies an increasing MOD price, escalating at the GRI.

Both costs and oil price assumptions are escalated to MOD for each year of the project. The cash flow calculation is carried out in MOD to yield an MOD net cash flow, as shown in Table 6.22.

Table 6.22 Project cash flow incorporating inflation assumptions.

Commercial and fiscal assumptions		Tax and Royalty terms	
Oil price	50	$/bbl	
Oil price escalation	3	% p.a.	
General rate inflation	5	% p.a.	
Fixed opex	5	% cumulative capex	
Variable opex	5	$/bbl	
Royalty rate	10	%	
Tax rate	60	%	
Capital allowance	25	% p.a. straight line	

YEAR			1	2	3	4	5	6	7	8	9	TOTAL
Flat RT Oil Price	50	$/bbl										
MOD oil Price			51.5	53.0	54.6	56.3	58.0	59.7	61.5	63.3	65.2	
ANNUAL PRODUCTION		MMbbl	0	0	10	20	20	15	10	5	1	80
GROSS REVENUE		$m			546.4	1125.5	1159.3	895.5	614.9	316.7	65.2	4658.3
PLATFORM CAPEX		$m	160	250								410
FACILITIES CAPEX		$m		200	150							350
DRILLING CAPEX		$m	40	50	50	40						180
TOTAL BYC CAPEX		$m	200	500	200	40						940
TOTAL MOD CAPEX		$m	210.0	551.3	231.5	48.6						1041.4
Cumulative MOD Capex			210.0	761.3	992.8	1041.4	1041.4	1041.4	1041.4	1041.4	1041.4	
Capital allowance - Year 1 capex		$m			52.5	52.5	52.5	52.5				210.0
Capital allowance - Year 2 capex		$m			137.8	137.8	137.8	137.8				551.3
Capital allowance - Year 3 capex		$m			57.9	57.9	57.9	57.9				231.5
Capital allowance - Year 4 capex		$m				12.2	12.2	12.2	12.2			48.6
TOTAL CAPITAL ALLOWANCE		$m			248.2	260.3	260.3	260.3	12.2			1041.4
Unit Variable opex	5	$/bbl										
VARIABLE MOD OPEX		$m			57.9	121.6	127.6	100.5	70.4	36.9	7.8	514.9
FIXED MOD OPEX		$m			49.6	52.1	52.1	52.1	52.1	52.1	52.1	310.0
TOTAL MOD OPEX		$m			107.5	173.6	179.7	152.6	122.4	89.0	59.8	824.8
Royalty Rate	10	%										
ROYALTY		$m	0	0	54.6	112.6	115.9	89.6	61.5	31.7	6.5	465.8
TAX DEDUCTIONS		$m	0	0	410.4	546.5	556.0	502.5	196.1	120.7	66.4	2332.1
TAXABLE INCOME		$m	0	0	136.0	579.0	603.3	393.1	418.9	196.0	−1.1	2326.2
Taxrate	60	%										
TAX		$m	0	0	81.6	347.4	362.0	235.8	251.3	117.6	−0.7	1395.7
GROSS REVENUE		$m	0	0	546.4	1125.5	1159.3	895.5	614.9	316.7	65.2	4658.3
CAPEX + OPEX (TECHNICAL COST)		$m	210.0	551.3	339.0	222.2	179.7	152.6	122.4	89.0	59.8	1866.2
GOVERNMENT TAKE		$m	0	0	136.2	459.9	477.9	325.4	312.8	149.3	5.9	1861.6
MOD NET CASH FLOW		$m	−210.0	−551.3	71.1	443.3	501.7	417.6	179.7	78.4	−0.4	930.5
CUM. MOD NET CASH FLOW		$m	−210.0	−761.3	−690.2	−246.9	254.8	672.4	852.1	930.5	930.1	
RT NET CASH FLOW		$m	−200.0	−500.0	61.4	364.7	393.1	311.6	127.7	53.1	−0.3	611.6
CUM. RT NET CASH FLOW		$m	−200.0	−700.0	−638.6	−273.9	119.2	430.8	558.5	611.6	611.3	

In this calculation, the reference date for oil price and costs is at the beginning of Year 1. Oil price is assumed to be flat RT $50/bbl, increasing at 3% p.a., and MOD prices are calculated at end-year.

Capex and variable opex are assumed to escalate at the GRI of 5% p.a., and no market factor is applied. Fixed opex is a fixed percentage of the cumulative MOD capex. Care is taken not to simply escalate the sum of fixed plus variable BYC opex as this would double-count on the fixed opex, since cumulative MOD capex is already an escalated value.

The capital allowance is based on the straight line method of 25% per annum, applied to the MOD capex. This is reasonable in that the MOD capex is the sum actually spent in each year, and is deemed as the allowable expense. However, it may be argued that the capital allowance calculated in this way should itself be escalated over time using the inflation rate to preserve its RT value. This example does not escalate capital allowance, but in some fiscal systems, an uplift on capital allowance is applied to maintain the RT value. In the United Kingdom, for example, capital allowance uplift has been as high as 35% (multiplying the capital allowance in this example by 1.35) — this is more than adequate to compensate for the effect of inflation, and is in fact a concession or incentive in the fiscal system at the time.

Once MOD costs and revenue are escalated to MOD, the rest of the cash flow calculation including royalty and tax is performed in MOD. This is reasonable, as the government will be taking its share in each project year, as MOD. All MOD cash flow items are highlighted in pink in Table 6.22, finally resulting in the MOD net cash flow and the cumulative. Year 9 has again been excluded as beyond the economic lifetime, and decommissioning cost neglected.

The discount rate that would make the NPV of the MOD net cash flow equal to zero is the IRR, discussed in Section 6.6.

The next step is to deflate the MOD net cash flow to RT, representing the purchasing power of the cash flow to buy common goods. If the price of bread has escalated by the GRI throughout the project life, the RT values represent how many loaves of bread the project net cash flow can buy. The MOD-deflator step in Fig. 6.24 is the inverse of the escalation step but uses only the GRI, without any market factor.

$$\text{RT cash flow} = \text{MOD cash flow}/(1 + \text{inflation rate})^t$$

The RT net cash flow and cumulative are shown in blue in Table 6.22.

The discount rate that would make the NPV of the RT net cash flow equal to zero is known as the **real rate of return** (RROR). This can be calculated using the Excel function = IRR but applied to the row containing the RT net cash flow. Excel does not have a function called RROR, but the mathematics is exactly the same — iterating to find a discount rate making the NPV of the string of numbers equal to zero.

In this example, the IRR of the project is 26.6% while the RROR is 20.6% as shown in Table 6.23.

Table 6.23 Comparison of IRR and RROR.

YEAR		1	2	3	4	5	6	7	8
MOD NET CASH FLOW	$m	−210.0	−551.3	71.1	443.3	501.7	417.6	179.7	78.4
				Excel formula		=IRR(E52:L52)			26.6%
						note E52:L52 is the MOD net cash flow			IRR
YEAR		1	2	3	4	5	6	7	8
RT NET CASH FLOW	$m	−200.0	−500.0	61.4	364.7	393.1	311.6	127.7	53.1
				Excel formula		=IRR(E55:L55)			20.6%
									RROR
NPV(8) CALCULATION				Excel formula		=NPV(8%,E55:L55)*(1+0.08)^0.5			280.6
						note E55:L55 is the RT net cash flow			NPV(8) $m

As an approximation, for low GRIs and small market factors, the difference between the IRR and RROR is approximately the GRI. Note that there is non-linearity when inflation is introduced, such as the treatment of capital allowance, so this approximation should be treated as such.

When inflation is introduced into the example, the IRR appears to have increased compared to the non-inflationary version. This should not be interpreted as a general effect, as the result depends upon assumptions about oil price escalation as well as inflation (and market factors, is used). For interest, if the above example is run with 5% GRI and no escalation in oil price, the IRR becomes 21.2% and RROR 15.4%. With no oil price escalation the effect of inflationary costs is to reduce the IRR and RROR of the project.

The final step in Fig. 6.24 involves discounting the RT net cash flow using the cost of capital. This straightforward discounting step is performed using Excel and is included at the bottom of Table 6.23. The correction for mid-year discounting has been applied and the NPV(8) of the project is $280.6 m.

The rigorous method followed in Fig. 6.24 allows the calculation of RROR from the RT net cash flow as well as the IRR from the MOD net cash flow. The cumulative RT net cash flow reflects the purchasing power of the net cash flow, making it easier to compare projects in different regions where inflation rates and market factors differ.

However, many companies choose not to include the step to calculate the RT net cash flow (and hence RROR), and instead move from the MOD net cash flow to NPV with one procedure. In choosing to do this, the deflation and discounting steps must be combined into one procedure, using an effective discount rate (r_e) which has been termed below as the '**combined discount rate**'. In the example, the combined discount rate (r_e) is calculated from

$$1 + r_e = (1 + r) \times (1 + i)$$

where
r_e = combined discount rate (%)
r = cost of capital (%)
i = general rate of inflation (%)
so in the example

$$1 + r_e = (1 + 0.08) \times (1 + 0.05) = 1.134$$
$$r_e = 0.134 \text{ or } 13.4\%$$

At low rates of inflation (say less than 10% p.a.), a reasonable approximation is to add the cost of capital and the GRI and to use this as the effective discount rate. This is not mathematically correct, but is a working solution. Table 6.24 shows the error that this introduces to the NPV calculation. Applying the effective discount rate of 13.4% yields the correct NPV(8) of $280.6 m, while the approximation using 13.0% yields $293.4 m.

Table 6.24 Effective discount rate combining cost of capital and deflation.

YEAR		1	2	3	4	5	6	7	8
MOD NET CASH FLOW	$m	-210.0	-551.3	71.1	443.3	501.7	417.6	179.7	78.4
APPLY EFFECTIVE DISCOUNT RATE 13.4%			Excel formula		=NPV(13.4%, E52:L52)*(1+0.08)^0.5				280.6
									NPV
APPLY APPROXIMATE DISCOUNT RATE 13.0%			Excel formula		=NPV(13.0%, E52:L52)*(1+0.08^)0.5				293.4
									NPV

The main question to ask when discounting a net cash flow is 'what is being discounted — an MOD or RT net cash flow?'. If cost and price have been escalated to MOD, then discounting the MOD net cash flow amounts to deflating to RT and discounting to reflect the cost of capital. A company that discounts at 12% is most likely to be combining the deflation and discounting step shown in Fig. 6.24, with a cost of capital of 8% and an inflation rate of 4%. The same project would discount the RT cash flow at the cost of capital 8% and get the same NPV. This is an area of great confusion between partners in the industry, and the question is worth asking to provide clarity of the method used.

A second important point to clarify is whether the rate of return being quoted is an IRR or RROR. Remember that these will differ by approximately the rate of inflation in most cases (where inflation levels are low). Again, the method is important, since different parties may present the same project as a rate of return, one of which is an IRR and the other an RROR. The IRR will always appear more attractive, assuming inflation is positive. The opposite of inflation is of course deflation, which is less common in most of the world, but has been a significant feature in the Japanese economy.

Through this section, two more project indicators have been introduced: IRR and RROR. The following chapter will add further project economic indicators based on the discounted cash flow technique and then summarise the measures by which the attractiveness of a project may be judged.

References

[1] International Energy Agency, World Energy Outlook 2019.
[2] US Energy Information Administration, International Energy Outlook 2014.
[3] Wood Mackenzie. https://www.woodmac.com/.
[4] International Petroleum Exchange. https://www.theice.com/publicdocs/futures/Options_User_Guide.pdf.
[5] International Monetary Fund (IMF), World Economic Outlook (WEO). https://www.imf.org/external/pubs/ft/weo/data/assump.htm.
[6] S&P Global Platts. https://www.spglobal.com/platts/en/commodities/natural-gas.
[7] German Federal Office of Economics and Export Control. https://www.bafa.de/EN/Foreign_Trade/Export_Control.
[8] ICIS Heren Energy Limited. https://www.icis.com/explore/commodities/energy/gas/.
[9] Energy Intelligence Group, Natural Gas Week. http://www.energyintel.com/pages/about_ngw_datasource.aspx.
[10] IHS Group, IHS QUE$TOR cost estimating service. https://cdn.ihs.com/www/questor/Quick-Start-Guide.pdf.
[11] Nomitech Ltd, CostOS estimation. https://www.nomitech.eu/company#.
[12] AGR, P1 software. https://www.agr.com/our-capabilities/software/p1.
[13] IHS Markit, Upstream Capital Costs Index (UCCI) and Upstream Operating Costs Index (UOCI). https://ihsmarkit.com/products/upstream-costs-expenditures.html.
[14] IHS Markit, Rushmore Drilling Performance Review. https://ihsmarkit.com/products/drilling-performance-review.html.
[15] McKinsey, The Art of Project Leadership: Delivering the World's Largest Projects, September 2017. https://www.mckinsey.com/media/McKinsey/Industries/Capital-Projects-and-Infrastructure/.
[16] 8over8, White Paper, Capex Project Overruns: Tackling the Lasting Negative Effects of Risk, January 2016. www.8over8.com.
[17] Rystad Energy, Rystad Energy UCube. https://www.rystadenergy.com/newsevents/news/press-releases/Rystad-Energy-ranks-the-cheapest-sources-of-supply-in-the-oil-industry-/.

[18] D. Johnston, International Petroleum Fiscal Systems and Production Sharing Contracts, PennWell, 1994.
[19] EY, Global oil and gas tax guide. https://www.ey.com/Publication/vwLUAssets/ey-global-oil-and-gas-tax-guide-2019/$FILE/ey-global-oil-and-gas-tax-guide-2019.pdf.
[20] Deloitte, International Oil & Gas Tax Guides. https://www2.deloitte.com/global/en/pages/energy-and-resources/articles/international-oil-gas-tax-guides.html.
[21] D.E. Smith, Ring Fence Provisions — How Important Are They in Determining Value of Petroleum Projects in a Portfolio, SPE-94631-MS, 2005, https://doi.org/10.2118/94631-MS.
[22] R.K. Elmaci, International Energy Negotiations; Production Sharing Contract Process and Cross Border Pipeline Bargaining, February 2019. http://etd.lib.metu.edu.tr/upload/12623070/index.pdf.
[23] T. Bramono, D.R. Galih, Policy shifting towards new regime of production sharing contract, in: Conference Proceedings, GeoBaikal 2018, vol. 208, EAGE, August 2018.
[24] https://www.wallstreetmojo.com/weighted-average-cost-capital-wacc/.

CHAPTER 7

Economic indicators from the DCF

Contents

7.1 Indicators from the cash flow and discounted cash flow	207
7.2 Indicators of efficiency	208
7.2.1 Capital efficiency	208
7.2.2 Technical cost efficiency	210
7.2.3 Unit technical cost and break-even price	211
7.2.4 The interpretation and application of IRR	214
7.2.5 Cautionary use of Excel = IRR function	220
7.2.6 IRR and RROR as a hurdle rate	221
7.3 Summary of economic indicators	221
7.4 Incorporating decommissioning cost	224
7.5 Choosing between alternative projects	225
7.6 Distinguishing net cash flow from net income	228
References	229

7.1 Indicators from the cash flow and discounted cash flow

From developing the net cash flow and cumulative net cash flow, and by applying the discounted cash flow (DCF) technique, the indicators introduced so far are summarised in Table 7.1.

Table 7.1 Economic indicators from net cash flow and discounted cash flow.

Economic indicator	Unit	Source
Pay-as-you-go	year	Net cash flow
Economic lifetime	year	Net cash flow
Ultimate cash surplus (undiscounted)	$	Cumulative net cash flow
Payback time	year	Cumulative net cash flow
Maximum exposure	$	Cumulative net cash flow
Net Present Value (NPV)	$	Discounted net cash flow
Internal rate of return (IRR)	%	Money of the Day (MOD) net cash flow
Real rate of return (RROR)	%	Real Terms (RT) net cash flow

These basic indicators are detailed in Chapter 6. However, with the exception of the rate of return metrics, they do not inform the investor about how efficiently the capital investment is utilised. This is an important aspect in selecting between projects when multiple opportunities are available. Remember that there is always the option to make a risk-free investment (e.g. guaranteed government bonds), or indeed to do nothing.

7.2 Indicators of efficiency

7.2.1 Capital efficiency

Imagine a choice between two projects, one that delivers NPV(8) $100 m and a second that delivers NPV(8) $200 m. The second would seem to be more attractive, since it creates more value, and companies are in business to create value, which is readily measured using NPV. However, this is only partial information about the attractiveness of the project, and other aspects should be considered. Table 7.2 provides the detail of the capex required in each of these two projects, along with the ratio of NPV per unit of capex invested.

Table 7.2 Capital efficiency ratio.

Opportunity	NPV(8) ($m)	Capex ($m)	NPV(8)/Capex (−)
Project A	200	600	0.33
Project B	100	200	0.50

The ratio of NPV to capex is the capital efficiency of the project. In general

$$\text{Capital efficiency ratio} = \text{NPV}/\text{Capex}$$

This ratio goes by a number of names, including
- Profit-to-Investment Ratio (PIR)
- Profitability Index (PI)

Table 7.2 shows that it is more efficient to invest in Project B. With access to say $600 m of capital, it would be more efficient to invest in three Project B opportunities and deliver $300 m NPV than to invest in one Project A and deliver $200 m NPV. This assumes that three Project B opportunities exist.

In the unlikely case that the opportunity set is limited to just one Project A and one Project B, with $600 m available, then Project A would be favoured. In the more common case that capital is constrained and a range of investment opportunities exist, the PIR is a very useful way of allocating the limited capex across the projects available.

As a guideline, a PIR target for a major capital project is 0.4. However, any efficiency ratio with a value less than 1.0 is often viewed with suspicion. It is important to realise that the NPV in the numerator is a **net** present value and includes all costs (capex, opex, government take) and discounting has been applied to reflect the cost of capital, so that in principle any project with a positive NPV and thus a PIR greater than zero is attractive. In practice, a value of 0.4 is a guide to the minimum target value for an interesting major capital project. To calm the nerves of the suspicious observer, some companies choose to define the capital efficiency ratio by adding 1.0 to the definition above, so that

- alternative PIR definition $\text{PIR}_{alt} = 1.0 + \text{NPV}/\text{Capex}$

and the typical target for PIR_{alt} would be 1.4. Nevertheless, it may be referred to simply as the 'PIR'. This raises an important point about comparing profitability indices between different companies, being that the definition is crucial. It is advisable to state the definition, rather than simply use the term. One company's $\text{PIR} = 0.4$, may be another company's 1.4.

Another example of variation in the definition of this indicator is to discount the capex in the denominator at the cost of capital when calculating (NPV/PV Capex). This is often called the Discounted Profitability Index (DPI).

- Discounted Profitability Index(DPI) = NPV/PV Capex

Again, some companies add 1.0 to this indicator, so it is advisable to state the definition.

The DPI version of the capital efficiency ratio is the most consistent, since all cash flows are discounted to a common reference date at the same discount rate. Discounting the capex is a matter of choice, but consistency within the company is most important. Since the capex is usually front-end loaded in a project, the discounting does not make a significant difference to the resulting ratio.

7.2.2 Technical cost efficiency

The PIR, or equivalent ratio, focusses only on the efficient use of the capex, which is usually a front-end-loaded expenditure. Over the lifetime of a major capital project, the total undiscounted opex is usually similar to the total undiscounted capex, as noted in the examples in Chapter 6. The technical cost efficiency ratio presented in the last row of Table 7.3 is not commonly used in industry, but is recommended since it accounts for all technical costs throughout the lifecycle. In this book it is referred to as the Discounted Profit to Total Investment ratio (DPTI), and all cash flows are discounted at the cost of capital to determine the project NPV and PV costs.

$$\text{Discounted Profit to Total Investment(DPTI)} = \text{NPV}/(\text{PV Capex} + \text{PV Opex})$$

Consider two alternative development concepts for an oil field. Project P requires a fixed production platform and export pipeline, with large up-front capex, and lifetime opex. Project S is a sub-sea tie-back solution with limited capex for sub-sea equipment but high lifetime opex to pay for processing on a host platform an subsequent export.

Table 7.3 Efficiency ratio incorporating lifetime costs.

Discounted values	Unit	Project P (platform)	Project S (subsea)
NPV	$m	300	300
PV Capex	$m	500	300
PV Opex	$m	400	900
DPI = NPV / PV Capex	-	0.60	1.0
DPTI = NPV / (PV Capex + PV Opex)	-	0.33	0.25

On the basis of NPV both projects are equivalent. Using a traditional capital efficiency ratio such as DPI (= NPV/PV Capex) the tie-back option looks significantly more attractive. However, when taking into account the discounted opex, the platform solution appears better. Ignoring the lifetime opex can bias the decision, and hence the recommendation to also consider DPTI.

A typical target for this indicator is DTPI ≥ 0.2. Note that the impact of including the opex approximately halves the DPI target of ≥ 0.4 for major capital projects.

7.2.3 Unit technical cost and break-even price

Unit technical cost (UTC) considers only the technical cost of producing a barrel of oil, and in its simplest form

$$\text{Unit Technical Cost (UTC)} = (\text{Capex} + \text{Opex})/\text{total production} (\$/\text{bbl})$$

It can be refined by discounting the annual costs, and also discounting the annual production

$$\text{PV Unit Technical Cost (PV UTC)} = (\text{PV Capex} + \text{PV Opex})/\text{PV total production} (\$/\text{bbl})$$

It may seem confusing to discount a production profile to calculate PV production, since Chapter 5 introduced discounting as a method of calculating the PV of future cash flows. However, the relationship between production and gross revenue is simply oil price, which is often forecast as a single flat value (gross revenue = production × price). Discounting annual production to PV production is therefore analogous to PV revenue, and distinguishes between alternative development projects that deliver reserves on a long low profile or a high short profile.

All investments being equal, the early production case in Fig. 7.1 is preferable to the late production case, even though both deliver the same reserves. The PV production in green is higher than the PV production in red, so for the same PV capex and opex, the PV UTC of the green case would be lower and therefore preferable to the red case.

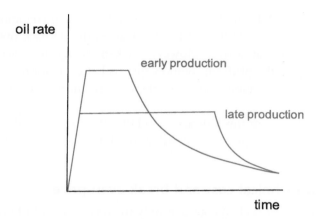

Figure 7.1 Significance of PV production.

UTC is simple to calculate, and is independent of the fiscal system — it has no tax or royalty element. It has several key uses.

Firstly, for any particular region and reservoir complexity, the UTC should be similar, and UTC therefore serves as check that the appropriate development concept is being proposed. Fig. 7.2 shows the average UTC in various countries (using a Statista source which reviewed over 15,000 oil fields across 20 countries, Ref. [1]).

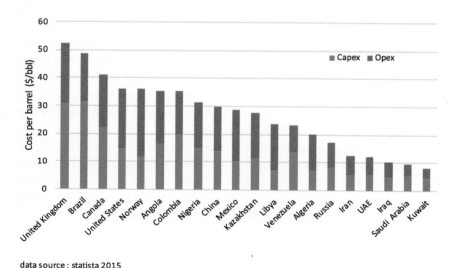

data source: statista 2015

Figure 7.2 Average unit technical costs on a country basis.

The countries selected in Fig. 7.2 represent the top 20 global oil producers. While these exact figures will change over time due to maturity in the region, technology enhancements and local labour costs, the contrasts are dramatic. As expected, the costs in Saudi Arabia are low, as a result of the low reservoir complexity, high well productivity, geography (onshore, warm climate) and proximity to infrastructure. It is not meaningful to compare the UTC of a development proposal in the North Sea with that in Saudi Arabia, but within the North Sea region, major capital projects should have similar UTCs to each other.

Secondly, UTC can be used as an indicator of the break-even oil price, itself an economic indicator. In this definition, the break-even oil price is the price required to cover the investment, but not make any profit, and should therefore be the same as the UTC. No tax calculation is required in the use of break-even price, since tax is only paid on profit. In Fig. 7.2, the average break-even oil price required for Norway would be $36/bbl. Note that if royalty is payable on gross revenue, the break-even price should be uplifted to UTC/(1.0 − royalty rate).

The simplicity of the UTC is appealing as a target to set for the project team, being easily understood and requiring little detailed knowledge of petroleum economics. Setting a goal to generate a field development plan that delivers production at a UTC less than say $40/bbl is clear to all. It implies that the project will make a pre-tax and royalty profit at an oil price above $40/bbl.

Thirdly, UTC can be used as a ranking tool when faced with the luxury of choosing which fields to develop from a portfolio of development opportunities. Imagine having a production constraint such as an export quota or a maximum export pipeline capacity, but also the possibility to produce above this constraint by developing all the opportunities available. There is a logic to ranking the options using UTC and start selecting in order of the fields with the lowest UTC. This should meet the constraint at the lowest average technical cost.

This can be effective in the short term, but may not optimally develop the portfolio in the long term. An unintended consequence of cherry-picking the lowest UTC fields is that it may not be consistent with the best use of infrastructure, and higher but still acceptable UTC developments may become stranded assets.

Another context for break-even price is shown in Fig. 7.3, based on 2020 data from Rystad Energy, a data services and consultancy company based in Norway (Ref. [2]).

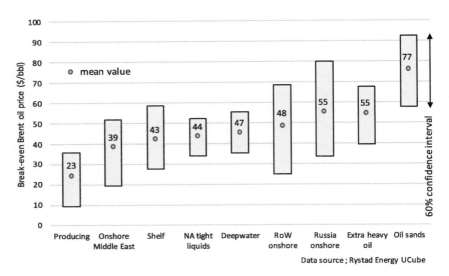

Figure 7.3 Break-even Brent oil price by region at 2020.

In Fig. 7.3 the break-even price refers to the oil price required to return zero NPV when the project net cash flow is discounted at 7.5%. The 'Producing' category represents the average of worldwide producing fields at the end of 2020. All other categories refer to non-producing fields at this date, meaning they are planned developments. NA tight liquids refers to North America shale oil, where break-even costs fell over the period 2010–20 due to the application of improved technology and batch drilling approaches. RoW refers to 'rest of world', being those regions outside the specific categories in the graph.

Again, this version of break-even price is useful for both regional benchmarking and testing the viability of development against the forecast Brent crude oil price.

7.2.4 The interpretation and application of IRR

Section 6.6 described the calculation of internal rate of return (IRR) and real rate of return (RROR). These are important economic indicators, but must be calculated and interpreted carefully. The IRR is the discount rate which, when applied to the MOD net cash flow, yields a zero NPV. Recall the PV profile shown in Fig. 6.22. If the IRR exceeds the weighted average cost of capital (WACC), then the project NPV must be positive when discounting the net cash flow at the cost of capital. Fig. 7.4 shows the PV profile in which the IRR is 25.4% and indicates the project NPV(8) as $406 m, assuming a WACC of 8%.

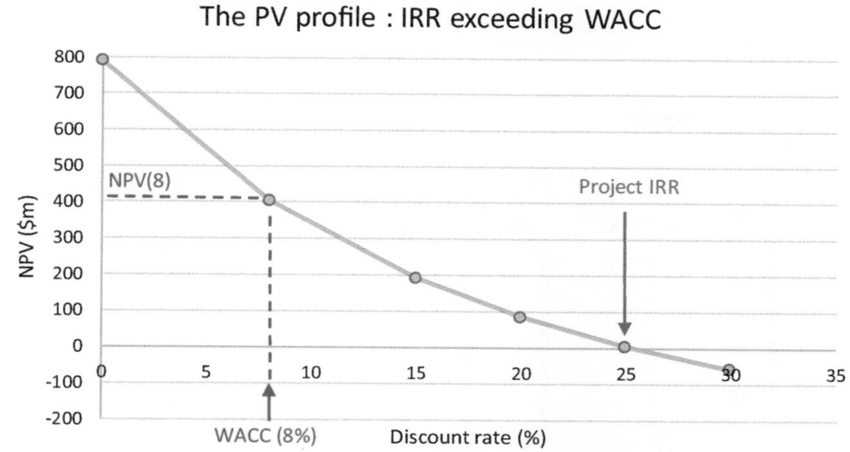

The PV profile : IRR exceeding WACC

Figure 7.4 PV Profile showing IRR and NPV at the cost of capital.

The simple analogy is that the project is comparable to a bank offering 25.4% p.a. interest rate on an investment. If one can borrow money at a cost of 8%, then investing in the project appears very attractive as the margin between the cost of borrowing and the promised rate of return is large. The narrower the margin between the known cost of capital and the IRR, the less attractive the project. This raises the question of the minimum acceptable IRR for a project to remain attractive.

A target IRR for a project can be set, and referred to as a **screening rate** or **hurdle rate of return**. This is the minimum acceptable IRR for a project to progress to further consideration for investment. For E&P companies, a typical hurdle rate for IRR is in the range 15%–25%, and typical WACC is 7%–10%. The margin between IRR and WACC indicates the robustness of the project, meaning its ability to withstand erosion of IRR while still returning at least the cost of capital. Erosion may be due to the project exposure to risk events such as drop in oil price, technical cost overrun, reduction of reserves, delays to production or change of the fiscal rules.

Fig. 7.5 illustrates the use of IRR as a screening tool; a hurdle that the project economics must clear in order for the project to progress to become ranked against other investment opportunities.

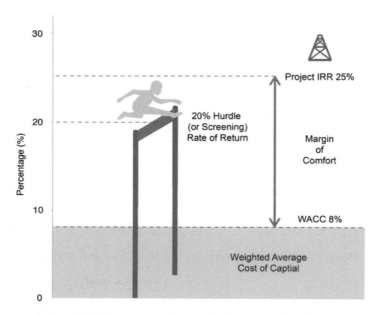

Figure 7.5 IRR as a screening tool with a target hurdle rate.

In principle, if the project economics indicate an IRR of, say 25% as in Fig. 7.5 and the project performs exactly to plan, the project should deliver an IRR of 25%. As indicated in Section 6.3, in the majority of cases, projects overrun the initial capex estimate, and fall behind the planned schedule. Some projects fail dramatically, such as the Yme project in Norway where the production platform was placed offshore in 2012, only then to have cracks detected in the legs of the structure, which was subsequently removed entirely and scrapped (Ref [3]).

Rarely is a review performed to determine the actual IRR of a project after it has run its course, and even more rarely will this be published. However, company annual reports provide information with which to calculate the return on average capital employed (ROACE). The ratio requires care in its definition. The average capital employed is the total capital employed in the business, which is all the company's assets (plant and equipment, cash, accounts receivable, inventory, stock) with all liabilities removed (short-term and long-term debt, tax and accounts payable), averaged over the reporting year.

The return in the numerator can be defined in several ways, such as the EBITDA (earnings before interest, tax, depreciation and amortisation) which represents the gross revenue less all costs (capex, opex, exploration costs, research and development costs, but not including interest payments, taxes, depreciation and amortisation). Since IRR is based on a post-tax net cash flow, but does not include any interest costs, the best comparison between IRR and ROACE would be to include tax but not interest in the numerator (Return). ExxonMobil have consistently presented a version of ROACE in their annual reports which includes tax but does not include interest payments in the numerator, and is therefore useful in broadly comparing project IRR with company ROACE.

Even with this definition, ROACE is the sum of all project performances within a specific year, while IRR is a project indicator based over the lifetime of the project. However, for a large company with many projects, there should be a broad relationship between the two. For example, if the company only approves projects with an IRR in excess say 20%, an annual ROACE performance over a number of years would be expected to be at least 20%, if all projects perform to expectation.

Data in Figs 7.6 and 7.7 are sourced from company annual reports and show ROACE over the period 2008−17 for several major IOCs.

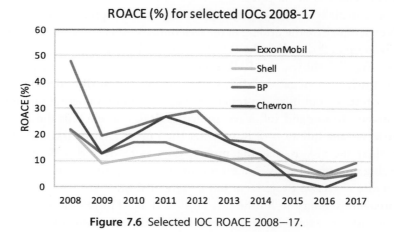

Figure 7.6 Selected IOC ROACE 2008−17.

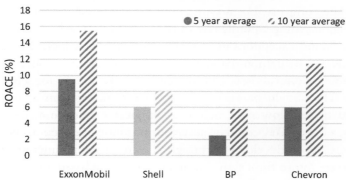

Figure 7.7 Selected IOC average ROACE over 5 and 10 years.

The companies shown are integrated oil companies with a downstream business, though the upstream E&P divisions dominate their overall financial results.

It has been noted that strictly speaking ROACE and IRR are not directly comparable. IRR is a project lifecycle metric and ROACE an annual accounting measure. Nevertheless, over longer term, ROACE would be expected to be close to the hurdle rate for IRR, which is typically in the range 15%—25% for major projects. Figs 7.6 and 7.7 indicate that this is not the case. The shortfall of ROACE compared to the hurdle IRR represents erosion of project value due to several factors.

The project cash flow focusses on the direct project costs and revenues, and does not include costs such as the exploration and appraisal activity, R&D costs, central office overhead cost allocations and other administrative costs, all of which erode overall project rate of return. ROACE reflects short-term dramatic changes in oil price. Fig. 7.6 demonstrates the impact of the oil price falls in 2008 and 2014. Project failures and catastrophic events such as the BP's experience on the drilling of the Macondo prospect with the Deepwater Horizon rig in 2010 affect the ROACE in the short term.

As Fig. 7.4 shows, the IRR must exceed the WACC, otherwise the project delivers a negative NPV. Given the erosion of IRR demonstrated, the margin of comfort illustrated in Fig. 7.5 must be great enough to withstand erosion and still leave an actual IRR that exceeds the WACC. This leaves the question of how to set the hurdle rate for IRR. If set too low then the risk is that the project does not cover the WACC after suffering erosion. However, if set too high then opportunity will be lost, as otherwise attractive projects would be screened out.

Imagine setting the hurdle rate at 40% IRR. This would exclude any typical major capital project, and limit the selection to short-term incremental projects such as those described in Chapter 9. A balance needs to be found for the hurdle rate of return. This becomes a matter of taste for the company, but is typically in the range 15%—25%.

Even after setting a hurdle rate of IRR, caution must be exercised in its application, and exceptions need to be made depending on the proposed project, as will be demonstrated.

Suppose a company sets a hurdle rate of IRR at 20% for screening of projects. This should be taken as a guideline, rather than a rule. For a project with very low risk, such as a compressor upgrade on an existing platform, a lower IRR could be accepted, assuming of course it still exceeds the WACC. For a development project in a country with high political risk or uncertain future fiscal terms, a higher hurdle rate could be justified to protect against erosion due to these factors.

A typical LNG plant construction project has an IRR of between 10% and 15%, so setting a hurdle rate of 20% would screen out such projects immediately. However, erosion of the LNG project IRR due to sales price uncertainty can be reduced by agreeing long-term fixed price gas sales contracts, making the typical IRR for LNG projects acceptable. In summary, the hurdle rate of IRR should be used in a flexible manner, treated as a guide rather than as a rule.

Fig. 7.8 shows the PV profile of two alternative projects for a company that has a guideline hurdle IRR of 20% and a WACC of 8%, and is facing a choice between investing either in Project A or Project B.

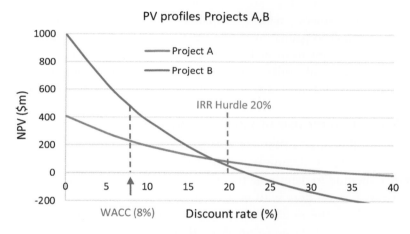

Figure 7.8 The IRR conundrum.

Both projects exceed the 20% IRR hurdle, and are therefore both of interest for further consideration. Based on IRR, Project A is the better (35% vs 22%). However, the NPV(8) of Project B is superior ($470 vs $230 m), and this presents a conundrum.

Companies would generally state their business objective as being to create value for their owners. An approach to solving this conundrum is to screen the projects based on an IRR hurdle (20% in this example), and then select from the surviving projects based on NPV, which is a clear statement of value. This is a simple two-step process of screening based on IRR and ranking based on NPV. Applied to the example, it would favour Project B even though it has the lower IRR.

A counter to this approach is to recognise that Project A is more efficient in returning cash to the business — its rate of return of the investment is higher, and it probably has a shorter payback time than Project B. The classical economist John Maynard Keynes, who favoured selection based on IRR, would have recommended taking on multiple Project As. This assumes that multiple Project As are available. It also assumes that the investor has enough capital to afford the option of investing in Project B; if this is not the case then Project B is not a realistic option.

The main advice in addressing this conundrum is not to make a choice between projects based on IRR alone. A high IRR may represent a project with short payback and efficient return of cash, but the project may also deliver relatively little value to the business, making it somewhat immaterial. The term 'materiality' refers to the size of the value delivered (NPV).

However, performing short payback, low capital exposure, high IRR projects may be the business model for some companies, which can make great success from this, provided they are able to organise many simultaneous projects while maintaining acceptable overheads. This requires an agile organisation and a large opportunity set, and low-cost operators can profit from such activities in regions with mature oil fields. Apache, along with other similar low-cost operators, has made this approach successful in mature field redevelopment in the UK North Sea.

7.2.5 Cautionary use of Excel = IRR function

The Excel function = IRR can be applied to a net cash flow forecast to calculate IRR, as demonstrated in Section 6.6. The coding uses an iterative method to determine the discount rate that makes the NPV equal to zero. For a cash flow with initial negative values followed by a series of positive values, the result is unique. However, where there are multiple changes of sign of the net cash flow, there can be multiple solutions to IRR. This will occur in net cash flows as shown in Fig. 7.9, which in this

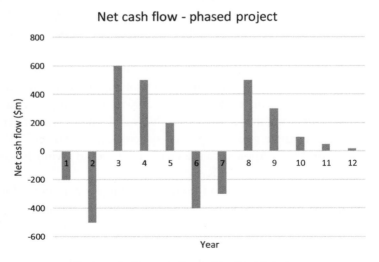

Figure 7.9 Net cash flow of a phased project.

case represents a staged project with an initial phase of development followed by further investment later, such as adding a stage of compression to a gas field or implementing an enhanced oil recovery scheme in a mature oil field.

With such irregular net cash flow profiles, the use of IRR is not recommended, as the solution that the = IRR function yields may be one of several values, despite the option to provide a guess for the solution in Excel as a seed for the iterative calculation.

7.2.6 IRR and RROR as a hurdle rate

The introduction to applying a hurdle rate of return for projects was made using IRR. For companies that discount the real terms net cash flow and calculate RROR as the discount rate yielding a zero NPV, the hurdle will most likely be set in terms of RROR. The use of RROR and cautionary comments apply in exactly the same way as discussed for IRR. Note, however, that a 20% RROR hurdle rate would be more stringent than a 20% IRR hurdle. Assuming a 3% general rate of inflation a 20% RROR hurdle rate would be approximately equivalent to a 23% IRR hurdle rate. Again, conversations with partners using different definitions of the hurdle rate can cause confusion unless the basis is clearly declared.

In summary, the IRR (or RROR) is an indicator both of robustness of the project, its ability to withstand erosion and still beat the cost of capital, and of efficiency, being the speed with which the project returns cash to the business. Projects with a high IRR will typically have a short payback time. As a crude rule of thumb, the payback time will be the reciprocal of the IRR — a project with a 20% IRR will have a payback time of approximately 5 years. This is a useful rule of thumb to check for calculation errors.

7.3 Summary of economic indicators

Table 7.4 illustrates the type of information provided by the economic indicators introduced in the last two chapters. The dark green symbols indicate the primary information, light green secondary. If pressed to select a single most important indicator, most company managers would choose NPV, on the basis that it presents value, and the business exists to generate value for its owners. The table shows that while this may be true, NPV is one-dimensional, as it does not inform the investor about any of the other aspects.

Table 7.4 Information from economic indicators.

Economic indicator	Unit	Value	Timing	Risk	Efficiency
Pay-as-you-go	year		●	○	○
Economic lifetime	year		●	○	
Ultimate cash surplus (undiscounted)	$	○			
Payback time	year		●	●	○
Maximum exposure	$			●	
Net Present Value (NPV)	$	●			
Internal rate of return (IRR)	%			●	●
Real rate of return (RROR)	%			●	●
Unit Technical Cost (UTC)	$/bbl			○	●
PV Unit Technical Cost (PV UTC)	$/bbl			○	●
Discounted Profitability Index (DPI)	ratio				●
Discounted Profit to Total Investment Index (DPTI)	ratio				●

To gain a fuller understanding of what the project offers, it is recommended that all of the above indicators should be presented, with the possible exception of pay-as-you-go time, which is less commonly used. Which of the indicators is most useful in selecting between alternative projects depends on the objectives and needs of the business and its current situation.

Table 7.5 shows a summary of the economic indicators for the project introduced in Table 6.22 in Chapter 6, which included inflation, but excludes decommissioning costs.

Table 7.5 Summary of indicators for tax and royalty example.

Commercial and fiscal assumptions		Tax and Royalty terms
Oil price	50	$/bbl
Oil price escalation	3	% p.a.
General rate inflation	5	% p.a.
Fixed opex	5	% cumulative capex
Variable opex	5	$/bbl
Royalty rate	10	%
Tax rate	60	%
Capital allowance	25	% p.a. straight line

YEAR			1	2	3	4	5	6	7	8	9	TOTAL
Flat RT Oil Price	50	$/bbl										
MOD oil Price			51.5	53.0	54.6	56.3	58.0	59.7	61.5	63.3	65.2	
ANNUAL PRODUCTION		MMbbl	0	0	10	20	20	15	10	5	1	80
GROSS REVENUE		$m			546.4	1125.5	1159.3	895.5	614.9	316.7	65.2	4658.3
PLATFORM CAPEX		$m	160	250								410
FACILITIES CAPEX		$m		200	150							350
DRILLING CAPEX		$m	40	50	50	40						180
TOTAL BYC CAPEX		$m	200	500	200	40						940
TOTAL MOD CAPEX		$m	210.0	551.3	231.5	48.6						1041.4
Cumulative MOD Capex			210.0	761.3	992.8	1041.4	1041.4	1041.4	1041.4	1041.4	1041.4	
Capital allowance - Year 1 capex		$m			52.5	52.5	52.5	52.5				210.0
Capital allowance - Year 2 capex		$m			137.8	137.8	137.8	137.8				551.3
Capital allowance - Year 3 capex		$m			57.9	57.9	57.9	57.9				231.5
Capital allowance - Year 4 capex		$m				12.2	12.2	12.2	12.2			48.6
TOTAL CAPITAL ALLOWANCE		$m			248.2	260.3	260.3	260.3	12.2			1041.4
Unit Variable opex	5	$/bbl										
VARIABLE MOD OPEX		$m			57.9	121.6	127.6	100.5	70.4	36.9	7.8	514.9
FIXED MOD OPEX		$m			49.6	52.1	52.1	52.1	52.1	52.1	52.1	310.0
TOTAL MOD OPEX		$m			107.5	173.6	179.7	152.6	122.4	89.0	59.8	824.8
Royalty Rate	10	%										
ROYALTY		$m	0	0	54.6	112.6	115.9	89.6	61.5	31.7	6.5	465.8
TAX DEDUCTIONS		$m	0	0	410.4	546.5	556.0	502.5	196.1	120.7	66.4	2332.1
TAXABLE INCOME		$m	0	0	136.0	579.0	603.3	393.1	418.9	196.0	-1.1	2326.2
Taxrate	60	%										
TAX		$m	0	0	81.6	347.4	362.0	235.8	251.3	117.6	-0.7	1395.7
GROSS REVENUE		$m	0	0	546.4	1125.5	1159.3	895.5	614.9	316.7	65.2	4658.3
CAPEX + OPEX (TECHNICAL COST)		$m	210.0	551.3	339.0	222.2	179.7	152.6	122.4	89.0	59.8	1866.2
GOVERNMENT TAKE		$m	0	0.0	136.2	459.9	477.9	325.4	312.8	149.3	5.9	1861.6
MOD NET CASH FLOW		$m	-210.0	-551.3	71.1	443.3	501.7	417.6	179.7	78.4	-0.4	930.5
CUM. MOD NET CASH FLOW		$m	-210.0	-761.3	-690.2	-246.9	254.8	672.4	852.1	930.5	930.1	
RT NET CASH FLOW		$m	-200.0	-500.0	61.4	364.7	393.1	311.6	127.7	53.1	-0.3	611.6
CUM. RT NET CASH FLOW		$m	-200.0	-700.0	-638.6	-273.9	119.2	430.8	558.5	611.6	611.3	

Indicator	Value	Unit	Definition
cumulative net cashflow	930.5	$mln	undiscounted cumulative net cashflow
pay-as-you-go	3	yrs	net cashflow first positive
payback time	4.5	yrs	cumulative net CF returns to zero
maximum exposure	551.3	$mln	minimum of cumulative net cashflow
NPV(8)	280.6	$mln	net present value (mid-year discounting)
IRR	26.6	%	internal rate of return
RROR	20.6	%	real rate of return
UTC	22.1	$/bbl	capex plus opex per barrel
PV UTC	26.5	$/bbl	PV (capex plus opex) per PV barrel
NPV / capex	0.30	-	NPV per unit of capex
NPV / PV Capex	0.35	-	NPV per unit of PV capex
NPV / PV(Capex + Opex)	0.23	-	NPV per unit of PV (capex plus opex)

7.4 Incorporating decommissioning cost

The project cash flows introduced so far have ignored the impact of decommissioning cost, which will be incurred when the project net cash flow first becomes permanently negative, at the economic lifetime. Decommissioning cost can be significant, depending on the complexity of the project construction and the location. The money of the day (MOD) cost of decommissioning sub-sea wells can be as costly as the original MOD drilling and completion cost.

The fiscal treatment of the decommissioning cost will vary between host countries. In a concessionary regime, the decommissioning cost may be an allowable expense against previously paid taxes, effectively creating a refund from the host government and potentially significantly reducing the net of tax cost to the operator. In a PSC regime, it may be a requirement to allocate part of the annual revenue to an abandonment fund, accruing over the producing lifetime of the project, and used at the time of need.

Table 7.6 illustrates the impact of including decommissioning cost in the tax and royalty example. It is assumed that decommissioning cost is incurred in Year 9, at 50% of the original development capex (in MOD) and is also 50% tax deductible against previously paid taxes. This cost is entered into Year 9 as a negative net cash flow ($260.3m). In reality, decommissioning cost is the sum of many activities, spread over a number of years.

Table 7.6 Including estimated decommissioning costs in the tax and royalty example.

Commercial and fiscal assumptions Tax & Royalty terms		
Oil price	50	$/bbl
Oil price escalation	3	% p.a.
General rate inflation	5	% p.a.
Fixed opex	5	% cumulative capex
Variable opex	5	$/bbl
Royalty rate	10	%
Tax rate	60	%
Capital allowance	25	% p.a. straight line

YEAR		1	2	3	4	5	6	7	8	9	TOTAL
GROSS REVENUE	$m	0	0	546.4	1125.5	1159.3	895.5	614.9	316.7	0.0	4658.3
CAPEX+OPEX (TECHNICAL COST)	$m	210.0	551.3	339.0	222.2	179.7	152.6	122.4	89.0	0.0	1866.2
GOVERNMENT TAKE	$m	0	0.0	136.2	459.9	477.9	325.4	312.8	149.3	0.0	1861.6
MOD NET CASH FLOW incl DECOM	$m	-210.0	-551.3	71.1	443.3	501.7	417.6	179.7	78.4	-260.3	670.1
CUM. MOD NET CASH FLOW	$m	-210.0	-761.3	-690.2	-246.9	254.8	672.4	852.1	930.5	670.1	
RT NET CASH FLOW	$m	-200.0	-500.0	61.4	364.7	393.1	311.6	127.7	53.1	-167.8	611.6
CUM. RT NET CASH FLOW	$m	-200.0	-700.0	-638.6	-273.9	119.2	430.8	558.5	611.6	443.7	

Indicator	Value	Unit	Definition
cumulative net cashflow	670.1	$mln	undiscounted cumulative net cashflow
pay-as-you-go	3	yrs	net cashflow first positive
payback time	4.5	yrs	cumulative net CF returns to zero
maximum exposure	551.3	$mln	minimum of cumulative net cashflow
NPV(8)	193.4	$mln	net present value (mid-year discounting)
IRR	23.9	%	internal rate of return
RROR	18.0	%	real rate of return
UTC	22.1	$/bbl	capex plus opex per barrel
PV UTC	26.5	$/bbl	PV (capex plus opex) per PV barrel
NPV / capex	0.21	-	NPV per unit of capex
NPV / PV Capex	0.24	-	NPV per unit of PV RT capex
NPV / PV(Capex+Opex)	0.16	-	NPV per unit of PV (RT capex plus opex)

A comparison of the economic indicators in Table 7.5 and Table 7.6 shows that the impact of including decommissioning costs on NPV, IRR and RROR is significant and demonstrates the importance of including the estimate of decommissioning costs. In this example, the impact is somewhat exaggerated since the project is relatively short, and the PV decommissioning cost is not as suppressed by discounting at the cost of capital as it would be for a project with an economic lifetime of say 20 years.

Note that in this example, the UTC and related measures have not included the decommissioning cost as they normally refer to development cost per barrel. It would be justifiable, however, to include decommissioning cost in the capex terms.

7.5 Choosing between alternative projects

If a company has multiple investment opportunities, and unlimited capital to invest, then a simple method of selecting between projects would be to screen the projects based on the chosen IRR (or RROR) hurdle, and then rank the surviving projects based on the NPV using the cost of capital as the discount factor. This approach is known as **screening and ranking**. This should ensure that all projects have passed a robust test of IRR and are then prioritised based on the projects that add most value to the business. The only constraints on project execution may then be time and human resources, the latter being availability of staff of contractors. This would be a very privileged position to be in, as more likely there is some constraint on the business, typically investment capital available, or limit to the number of project opportunities.

In the more typical situation of being capital constrained, then a capital efficiency ratio such as the DPI or DPTI becomes appropriate (see Table 7.4). Again, projects may be screened using the IRR hurdle, and then ranked according to the highest capital efficiency.

In a capital-constrained situation, the maximum exposure also becomes important in making several checks. Firstly, is the company able to meet the maximum exposure of the project, or that of several combined projects? If not, then the company should consider the timing of expenditure by phasing the investment in the project or the sequencing of several projects, or to partnering, or raising further capital. It is always useful to consider whether the company could survive if the project was lost at the point of maximum exposure of the project — if not, then serious consideration should be given to stopping the project in its current form.

If constrained by production rate through export system or by a production quota such as imposed within OPEC members, then ranking based on the lowest UTC will provide a solution that satisfies the constraint while producing at the lowest cost. This may be valuable in the short term, but may mean that projects with reasonable UTC are side-lined in favour of lower cost projects, and may become stranded as infrastructure ages. Section 7.2 referred to the potential unintended consequences of prioritising the low UTC options.

If the company is in a cash flow crisis and under threat of liquidation, then it needs to generate cash quickly. Short payback projects, which typically have a high rate of return but individually low NPVs, will help to solve the short-term challenges. Major capital projects become deferred in favour of short payback incremental projects such as well recompletions, additional perforations and low-cost upgrades to improve efficiency of operations. Even major IOCs have had periods where payback has become the primary economic indicator.

In summary, the importance of the economic indicators will vary depending on the company's position, and it is therefore useful to calculate all of these indicators, which provide the decision-maker with the ability to judge multiple aspects of the project, rather than focussing only on the tempting but one-dimensional NPV.

Table 7.7 introduces two projects available for investment. Assume that the investor has $300 m of capital to invest, the company WACC is 8% and the selected IRR hurdle rate is 20%. Both projects pass the IRR hurdle rate of 20%, and can progress to the ranking based on NPV. Hovis becomes the preferred project, since it delivers the highest NPV (calculated using mid-year discounting), despite having the lower IRR and the longer payback. Basing the decision on IRR alone would not maximise the NPV from the options available.

Table 7.7 Selecting between projects Hovis and McDougal.

			Hovis		McDougal		Hovis plus McDougal	
		Year	Net CF ($m)	Cum Net CF ($m)	Net CF ($m)	Cum Net CF ($m)	Net CF ($m)	Cum Net CF ($m)
		1	−80	−80	−60	−60	−140	−140
		2	−150	−230	−120	−180	−270	−410
		3	100	−130	160	−20	260	−150
		4	120	−10	70	50	190	40
		5	80	70	50	100	130	170
		6	60	130	30	130	90	260
		7	50	180			50	310
		8	30	210			30	340
		9						
Payback	(yr)		4.1		3.3		3.8	
Max exposure	($m)		−230		−180		−410	
NPV(8)	($m)		107		76		182	
IRR	(%)		24%		30%		26%	
NPV/max exposure	(−)		0.46		0.42		0.44	

Performing both projects simultaneously is not feasible since the combined maximum exposure of $410 exceeds the $300 m of capital available, as the final columns show. However, both projects are attractive, easily passing the IRR hurdle, and the investor should seek a method of performing both projects with the limited capital budget by phasing the projects. Table 7.8 shows options to phase the projects.

Table 7.8 Phasing of projects to remain within a capital constraint.

		Hovis first				McDougal first			
Year		Hovis Net CF ($m)	McDougal Net CF ($m)	Total Net CF ($m)	Cum Net CF ($m)	Hovis Net CF ($m)	McDougal Net CF ($m)	Total Net CF ($m)	Cum Net CF ($m)
1		−80		−80	−80		−60	−60	−60
2		−150		−150	−230	−80	−120	−200	−260
3		100	−60	40	−190	−150	160	10	−250
4		120	−120	0	−190	100	70	170	−80
5		80	160	240	50	120	50	170	90
6		60	70	130	180	80	30	110	200
7		50	50	100	280	60		60	260
8		30	30	60	340	50		50	310
9						30		30	340
Payback	(yr)			4.8				4.5	
Max exposure	($m)			−230				−260	
NPV(8)	($m)			172				175	
IRR	(%)			26%				27%	
NPV/max exposure	(−)			0.75				0.67	

The maximum exposure lies within the capital constraint of $300 m if either McDougal is delayed by 2 years or Hovis is delayed by 1 year. The choice between these two then depends on the company's key criteria, but the latter has a marginally favourable NPV, IRR and payback time. Since they are so close, other considerations such as logistics or production profile may be the deciding factor, but in either case, the limited capital constraint has been satisfied.

An additional indicator has been introduced in these tables: NPV/maximum exposure. If maximum exposure is of particular concern, as in this example, then using it as the denominator with NPV as the numerator shows the best use of the constrained item.

As a general principle of creating efficiency ratios, the prize (NPV) is divided by the constraint (maximum exposure in this case). Some companies use this indicator, which can also be refined by defining the denominator as the sum of the discounted annual net cash flows up to the point of maximum exposure.

7.6 Distinguishing net cash flow from net income

The focus of petroleum economics is on generating the project net cash flow and deriving the indicators summarised in this chapter to support investment decisions. At the same time, the accountant will be reporting or forecasting annual performance using different indicators such as the ROACE introduced in Section 7.2, and annual net income.

Project net cash flow is based on exactly when the cash flow components are expected to arise. It specifies the year in which sums of money are received or spent, and is performed on a project-by-project basis. In a tax and royalty system, for example,

$$\text{net cash flow} = \text{revenue} - \text{expenditure}$$
$$= \text{revenue} - \text{royalty} - \text{opex} - \text{tax} - \text{capex}$$

By contrast, the accountant is usually reporting or forecasting the performance of a portfolio of projects on a company-wide basis, and uses net income as a measure. Net income is also called profit or earnings. A fundamental principle of accounting is that of 'matching' the revenues earned in a year to the costs incurred in generating those revenues.

Many of the costs of generating the annual revenues will occur in the same year. However, there are significant exceptions, including depreciation of capital costs and provisions for future costs such as the eventual decommissioning cost.

Depreciation is an accounting term, not to be confused with the capital allowances introduced in Section 6.4.2. When a capital expenditure item is purchased, its value is entered on the company's balance sheet as a fixed asset at the value of its cost. This is called the original book value. Once in service, the asset value (book value) is reduced each year by a fraction, usually related to the expected life of the asset or the annual throughput compared to lifetime throughput (similar to the Unit of Production method introduced in Section 6.4.2). The annual sum by which the value is reduced is the annual depreciation, and this becomes a charge on the company's profit and loss account in that year. While depreciation is not a cash flow item, the accountant deducts this (rather than capex) from revenue in calculating net income.

Similarly, the provision of a cost for the eventual decommissioning is not a cash flow item in the producing year, but is again considered as a cost by the accountant.

If the only items that did not occur in the same year as the associated revenues were capital depreciation and decommissioning provision, then for annual accounting purposes

$$\text{net income} = \text{revenue} - \text{royalty} - \text{opex} - \text{tax} - \text{capital depreciation} - \text{decommissioning provision}$$

Taxation may also not occur in the same calendar year as the corresponding revenue, as in many tax systems there is an element of advanced corporation tax, with the balance being paid in arrears. The sums paid in advance and arrears will form part of the annual net cash flow for the economist, but the accountant will need to match the taxation to the revenues by deducting only the tax assessments for each period.

The reporting of net income is regulated by accounting principles, which define in detail how non-cash expenses are accounted for. Examples include the International Financial Reporting Standards (IFRS) in the EU, and the Generally Accepted Accounting Principles (GAAP) in the United States.

References

[1] https://statista.com/statistics/597669/cost-breakdown-of-producing-one-barrel-of-oil-in-the-worlds-leading-oil-producing-countries/.
[2] https://www.rystadenergy.com/newsevents/news/press-releases/Rystad-Energy-ranks-the-cheapest-sources-of-supply-in-the-oil-industry-/.
[3] Offshore Energy.Com, Ill-Fated Yme Field Could Start Production in 2019. (Posted 20 January 2017).

CHAPTER 8

Risk analysis and decision-making for development

Contents

8.1 Development planning using a scenario approach	231
8.2 Describing subsurface uncertainty — reservoir realisations	233
8.3 Creating development concepts	240
8.4 Applying investment themes	242
8.5 Selecting the favoured concept	244
8.6 Risk management of the development concept	245
8.6.1 Spider diagrams	246
8.6.2 Bow Tie model	251
8.6.3 The risk matrix	256
8.6.4 The risk register	259
8.6.5 A financial adviser's view on risk	261
8.6.6 Utility	268
References	274

8.1 Development planning using a scenario approach

A field development plan (FDP) details the decisions made for reservoir management, well design, type of fluid processing facilities and the export system. Options for these components are investigated at the Assess stage of the Stage Gate process introduced in Section 5.1, where a wide range of possibilities should be considered. Remember that at the Assess stage we should be asking whether a wide enough range of options has been looked at. To progress from the Assess to the Select stage, at least one development option should be demonstrated to be feasible when tested against reservoir uncertainties, meaning that it should pass the basic economic criteria, and be acceptable from HSE and regulatory perspectives. At the Assess stage, the development options will have been considered at a high level, with cost estimates being accurate to approximately $\pm 40\%$, as described in Section 6.3, and reservoir uncertainty will be of a similar scale of confidence.

In working through the Assess and Select stages, many oil companies have applied a matrix approach to combining the facilities and export options with the reservoir uncertainty, as shown in Fig. 8.1.

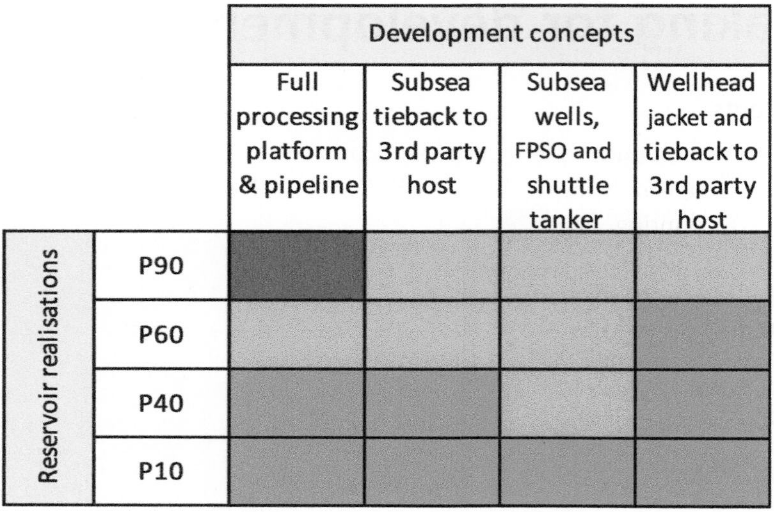

		Development concepts			
		Full processing platform & pipeline	Subsea tieback to 3rd party host	Subsea wells, FPSO and shuttle tanker	Wellhead jacket and tieback to 3rd party host
Reservoir realisations	P90				
	P60				
	P40				
	P10				

Figure 8.1 Matrix combining development concepts with reservoir realisations.

It is important to be clear about the nomenclature used in this matrix. A **development concept** is a discrete combination of choices for reservoir development, well planning, facilities and export system (e.g. water drive, horizontal wells, fixed steel platform and pipeline to shore). A **reservoir realisation** is a specific description of the reservoir geology, which will be a point within the range of reservoir uncertainty. In Fig. 8.1 several discrete reservoir realisations represent this range. While the company has control over the development concept, nature determines the reservoir properties; all the company can do is estimate a realistic range of reservoir uncertainty.

Table 8.1 Definitions for the scenario-based approach to development planning.

Term	Definition
Development concept	One choice of reservoir management plan, well type, production facilities and export system
Reservoir realisation	One description of the reservoir geology, chosen from within the range of reservoir uncertainty
Scenario	One combination of a particular development concept and one reservoir realisation

Each combination of a development concept and a particular reservoir realisation is called a **scenario**. Table 8.1 reinforces these definitions, as they can cause confusion if used loosely.

The traffic light colours represent how well each development concept meets the project screening criteria when exposed to each reservoir realisation. At the end of the Assess stage, at least one development concept should pass with mostly green lights in all reservoir realisations, and should not show any red lights, which would represent a project failure. At the Assess stage the criteria for the traffic lights will be relatively simple, such as meeting the screening rate of IRR, having a positive NPV when discounting at the cost of capital and meeting the HSE and regulatory standards.

Moving on to the Select stage of the process, more detail will be added to the description of reservoir uncertainty, and more scrutiny will be applied to the development concepts in terms of costing and technical robustness. During the Select stage, the company may introduce new development concepts or permutations of those proposed during Assess. The process is not necessarily linear, and iterations are quite acceptable in the pursuit of the optimal FDP. In Fig. 8.1, the best development concept appears to be the wellhead jacket with a tie-back to a host facility for processing and export. This may not be the concept finally selected, but has allowed the project to progress to the next stage of the Stage Gate process.

The next sections will describe the creation of reservoir realisations and development concepts.

8.2 Describing subsurface uncertainty — reservoir realisations

Of the technical elements of an FDP, the reservoir description usually carries the largest uncertainty. It is the role of the subsurface team to capture a realistic range of uncertainty, using the data available and the knowledge of the team.

As described in Chapter 4, reducing the range of uncertainty in reservoir description through appraisal can add significant value to the project by reducing the risk of over- or under-expenditure on facilities, but even after appraisal, a range of residual uncertainty will remain. Detailed reservoir modelling using computer-based simulation is not usually performed at the exploration or appraisal phases of the field life, but is generally a pre-requisite to gaining development approval from company management and from the host government.

When building reservoir models, it is tempting in terms of time and cost to create a base case model that represents a 'best guess', and then to run sensitivities around this to investigate the impact of varying input parameters on the resulting recoverable volumes and project value.

Fig. 8.2 (adapted from *Reservoir Model Design*, Ref. [1]) depicts the base case approach and illustrates that the sensitivities run are variations on the same geological concept. This method provides some comfort by generating a range of production profiles with associated values when run through the economic model, but the range of uncertainty is typically too narrow — nature often has a more vivid imagination that frequently creates reservoir outcomes that lie outside the range modelled. The drawback of the base case plus adjustment approach is that it centres on a solitary concept to describe the reservoir.

Focussing on a base case is a human bias identified by psychologists, who recognise that we prefer certainty to uncertainty. Therefore, when faced with trying to estimate a range of uncertainty in a parameter, our natural tendency is to think first of what we believe to be the correct answer and then create a range by moving the number up and down around the initial estimate. This, and many other pertinent human biases in dealing with uncertainty, is elegantly captured by Nobel Prize winning author Daniel Kahneman in his book *Thinking, Fast and Slow* (Ref. [2]). He names this particular bias 'anchoring and adjustment'.

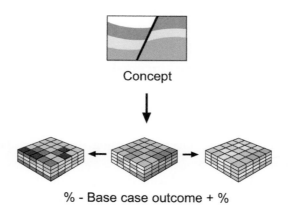

Figure 8.2 Base case reservoir modelling with adjustment to represent uncertainty range.

A statistical approach can be used to generate a range of reservoir models by honouring the limited data available from seismic, core and logs, and generating reservoir properties away from the control points according to given rules. This method still tends to centre on a single reservoir concept, and the range generated is typically still too narrow.

The Monte Carlo simulation method introduced in Section 4.2.2 provides a probabilistically derived range of uncertainty for hydrocarbons in-place. It should be recognised that the recovery factor used to estimate recoverable volumes is itself a function of the development plan (recovery mechanism, number and placement of wells, facilities constraints). Each reservoir realisation in the matrix in Fig. 8.1 should represent an in-place resource, and the recoverable volume for each scenario is then estimated by applying each development concept. In the matrix, four volumes have been used to represent the range of reservoir uncertainty (P90, P60, P40, P10); this is an arbitrary choice − more or less individual cases can be selected from the continuous distribution of in-place volumes.

A third approach to creating the reservoir realisations is to construct multiple discrete alternatives based on quite different, yet plausible, reservoir concepts that can provide a wider and more realistic range of uncertainty in the reservoir description. This method is referred to as multiple deterministic modelling, and is illustrated in Fig. 8.3, adapted from *Reservoir Model Design* (Ref. [1]), which is recommended reading on alternative modelling approaches.

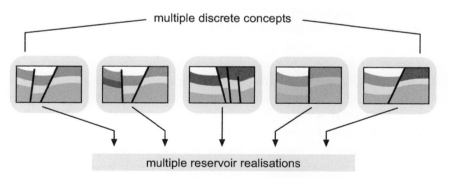

Figure 8.3 Multiple deterministic reservoir modelling.

Again, this is useful for the description of the static reservoir, that is, the in-place volumes. The recovery factor will be estimated by applying the reservoir development assumptions of a particular development concept (recovery mechanism, wells and facilities) in a dynamic model of each of these discrete cases. A dynamic model simulates fluid flow, well performance and facilities constraints, delivering a production profile and recoverable hydrocarbons.

A method of generating multiple deterministic models is to make combinations of the key parameters that the subsurface team consider to most strongly influence the range of plausible reservoir descriptions. The key parameters may be established by sensitivity analysis on simple models, such as cross-sectional models or material balance, or by using analogue fields and experience.

Suppose the key parameters are considered to be sand distribution (amalgamated or channelised sand units), oil-water contact depth (OWC — shallow or deep), fault sealing (faults open or sealing), aquifer strength (strong or weak) and permeability contrasts (homogeneous or heterogeneous), and that each parameter has a high and low estimated value. With five independent parameters and two values for each, there are 2^5 combinations. Building 32 deterministic models is ambitious; however, the range of outcomes may be represented by modelling just a selection of these cases.

Using a probability tree is a useful way to present all combinations, from which to select cases to take forward for modelling. Fig. 8.4 uses PrecisionTree to outline the combinations. The high and low for each parameter is each assigned a 50% probability. The lower half of the tree (channelised sands) has been collapsed to be able to fit the tree onto one page, which is a useful presentation feature of PrecisionTree. The probability of any particular outcome is shown on the leaves in blue. In this case each outcome is equally likely, the 32 outcomes each having a probability of 3.125%. The sum of all outcomes must of course add to 1.0, a useful check.

From the 32 outcomes, a selection is made to represent the overall range. The selection may be made by inspection or by using an experimental design technique which would select from the possible combinations just a limited set of cases that would nevertheless represent the range (Refs. [3,4]). Fig. 8.4 indicates four combinations selected from the upper half of the probability tree, the other four coming from the collapsed branches. Eight cases may then be taken forward for modelling.

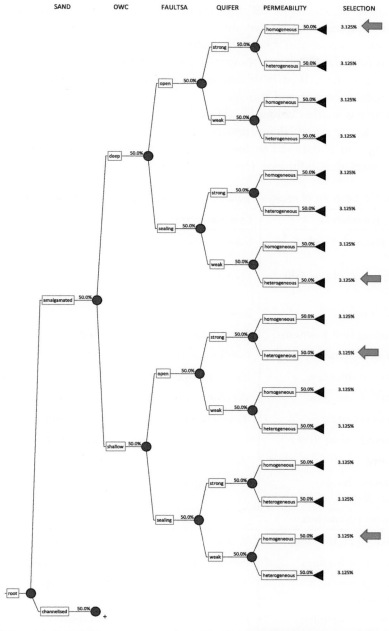

Figure 8.4 Probability tree approach to selecting combinations for modelling.

Another useful method for identifying the parameters that influence the range of plausible reservoir descriptions is to start with a list of general static and dynamic reservoir parameters and then consider each in terms of its range of uncertainty and its potential impact on production rates or recoverable volumes. The uncertainty range and potential impact can be traffic-lighted, with green representing low uncertainty range or low potential impact on recoverable volumes or production rate. Red is of course the opposite, and orange represents intermediate range and impact.

There is another colour to add to the traffic lights, blue. Blue represents 'don't know', and these parameters may be significant, but further investigation is required to find out. Table 8.2 is a suggested long list of parameters that influence recoverable volumes or production rates in a

Table 8.2 Long list of reservoir parameters that influence recoverable volumes.

Static parameters	Considerations
Top reservoir depth and extent	Seismic quality. Confidence in velocity models
Undrilled flanks or fault blocks	Hydrocarbon fill. Risking undrilled areas
Fluid contacts	Single or multiple contacts?
Bounding fault positions	Confidence in seismic migration
Intra-field faulting	Confidence in seismic, well test interpretation
Reservoir thickness	Variation away from well control
Reservoir net: gross	Variability. Missing thin reservoir units
Facies continuity	Reservoir pinch-out. Correlation from log data
Reservoir connectivity	Connectivity between reservoir units for sweep
Porosity	Variation away from control data (log, core)
Saturation–height relationships	Availability of core analysis, log data
Dynamic parameters	
Intra-field fault sealing	Will minor faults compartmentalise reservoir?
Presence of fractures	Presence. Influence on sweep and productivity
Heterogeneity of permeability	Thief zones, bypassed oil during sweep
Vertical permeability	Is development dependent on this?
Presence of shale baffles	Influence on sweep efficiency
Fluid properties	Single fluid type or variable?
Diagenesis	Have we sampled for this? Effect on production
Aquifer strength	Is aquifer likely to encroach during production?

field development context, split into static and dynamic categories. This list is not intended to be exhaustive and is a starting point for the traffic lighting process, which can be added to for any specific study. An additional column that is useful would note the root cause of the uncertainty. For example, if the connectivity of the channel sands is an uncertainty, this may be rooted in a lack of understanding of the environment of deposition — fluvial or deltaic. This could prompt further investigation using analogues.

The 'long list' may be considerably longer, and is a generic starting point for any subsurface review, recognising that some items on the list will not apply to some reservoirs.

Fig. 8.5 demonstrates a portion of the table with traffic light results.

Field Name		Misuri					Low / Moderate / High / Don't know		Low / Moderate / High / Don't know
Type of Uncertainty			General description of Uncertainty	What we know	Root cause	Range of Uncertainty	Impact on recoverable volume / production forecast		Impact of Uncertainty
GRV		Top structure / reservoir extent	Uncertainty in loop pick. Time-depth conversion. Limited well control on depth in flanks						
		Fluid contacts	Contact in wells poorly defined. Possible multiple contacts across the field						
		Fault position	Position of bounding fault and offset at intra-field faults						

Focus on the red/red and red/orange combinations
Don't forget blues – we don't know what we don't know

Figure 8.5 Example of the long list with traffic light results.

The long list can then be distilled down to a short list by focusing on those items that have both significant uncertainty and significant impact, prioritising those parameters that have red-red, red-orange or blue values. Remember that the unknown unknowns may be important, but require further investigation to find out. Fig. 8.5 includes columns in which to record what is already known, which should also state the supporting evidence, the root cause of the uncertainty and the potential impact on the recoverable volumes or on the production profile.

The resulting key uncertainties can become the input for the probability tree method shown in Fig. 8.4, or the subject of further work to determine whether they have genuine impact on the subsurface volumes or on production.

8.3 Creating development concepts

Development concepts can be broken down into categories covering

- subsurface (reservoir and wells)
- processing facilities
- disposal and export

The following tables (Tables 8.3–8.5) will illustrate some of the options for each of these categories. The example assumes an offshore oil field development, but project-specific tables would be adapted to the circumstances of the particular FDP. To clarify some of the less common terms, wet trees refer to wellheads on the seabed, dry trees to wellheads on a wellhead deck, SMART completions are tubular configurations that allow inflow from specific sections of the reservoir to be controlled remotely and an FPSO is a floating production storage and offloading vessel. A more detailed description of the options can be found in *Hydrocarbon Exploration and Production* (Ref. [5]).

Table 8.3 Development options for subsurface — reservoir and wells.

Category	Subsurface							
Element	Drive mechanism	Well placement	Well type		Completion type	Sand control	Artificial lift	EOR consideration
			Producers	Injectors				
Options	Depletion	Crestal producers	Vertical	Vertical	Commingled production	None	None	None
	Water injection	Crestal producers, downdip injectors	Deviated	Deviated	Zonal completion	Sand screens	Gas lift	Polymer flood
	Gas injection	Crestal injectors, downdip producers	Horizontal	Horizontal	SMART completion	Gravel packs	Electrical Submersible Pump (ESP)	Surfactant flood
	Combination	Pattern flood	Mixture	Mixture			Progressive Cavity Pump (PCP)	Combination

Table 8.4 Development options for facilities.

Category	Facilities						
Element	Wellheads	Number of drilling centres	Fluid gathering	Processing	Number of processing centres	Plateau oil rate	EOR capacity
Options	Dry trees	Single	Single site	Fixed jacket and topside processing	Single	High	None
	Wet trees	Multiple	Multiple sites	Semi-submersible with processing	Multiple	Medium	Polymer
	Combination		Subsea template	FPSO		Low	Surfactant
			Subsea manifolds	Wellhead jacket, third-party processing		Flexibility to expand	Combination

Table 8.5 Development options for disposal and export.

Category	Disposal and export			
Element	Excess gas handling	Produced water handling	Oil export	Gas export
Options	Reinject	Reinject	Dedicated pipeline to shore	Dedicated pipeline to shore
	Flare	Dispose to sea	Trunk line to third-party pipeline	Trunk line to third-party pipeline
	Offshore power generation		Storage vessel and shuttle tanker	None
			Offshore loading buoy	

There are many possible combinations of these options. Fig. 8.6 illustrates this point, considering just two elements of the development plan, the processing and export. It would be burdensome to consider all possible permutations of the options.

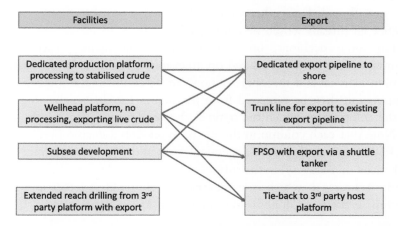

Figure 8.6 Example of plausible combinations of processing and export options.

In order to make the evaluation more manageable, it is useful to develop themes or criteria with which to string together choices that best satisfy each theme.

8.4 Applying investment themes

Themes are based on the ambitions of the investor, but limited by what is realistic given the abilities and constraints. The themes may represent broad strategies such as building a position of becoming
- leading operator of infrastructure in the region
- most effective user of leading technology
- providing tie-back opportunity for third-party business

The themes can be related to economic indicators summarised in Chapter 7, such as
- limiting the maximum exposure of the cumulative net cash flow
- becoming lowest unit technical cost operator in the region
- minimising payback time
- maximising recovery factor

The development options matrix is then used to indicate which options support the themes by threading across the subsurface, facilities and export components. Fig. 8.7 indicates this process using a condensed version of the elements, and threading through the options driven by maximising the oil recovery (in blue). The optimal choices for the other elements of the development plan are shown by colour, but not threaded together simply for clarity in the figure, but in practice all threads would be completed.

For some decisions, one option satisfies all themes, such as the requirement for sand screens and dry trees in the example. In this case, the decision is trivial and the single option can be considered as a given. Such columns can then be removed from the matrix for clarity, allowing focus on the columns for which the optimal choices diverge.

Note that each column in the matrix represents a decision required for the FDP. Some decisions are dependent, such as the consideration for EOR (enhanced oil recovery) in the subsurface element and the provision for this in the facilities, in which case they can be clustered together as a single column, as has been done in Fig. 8.7 under 'EOR capacity'.

Risk analysis and decision-making for development 243

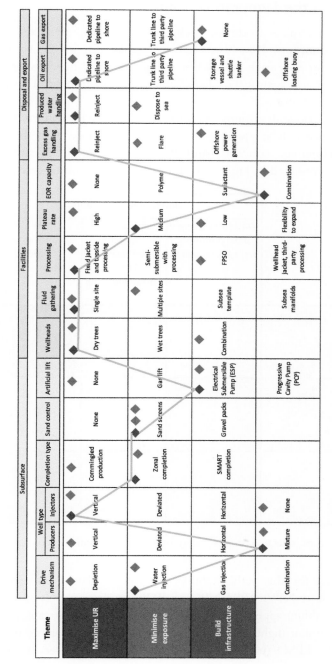

Figure 8.7 Threading across the development options matrix to satisfy a theme.

8.5 Selecting the favoured concept

The threads of choices made for each theme now represent alternative development concepts. Which of these is the best development concept now depends on how each fares when set against the range of subsurface realisations, and also against the overall criteria for investment. These tests will be considered in turn.

Testing alternative development concepts against the realisations uses the same matrix approach as presented in Fig. 8.1, but now with the more detailed development concepts, and the updated range of subsurface uncertainty, represented in Fig. 8.8 with multiple deterministic cases.

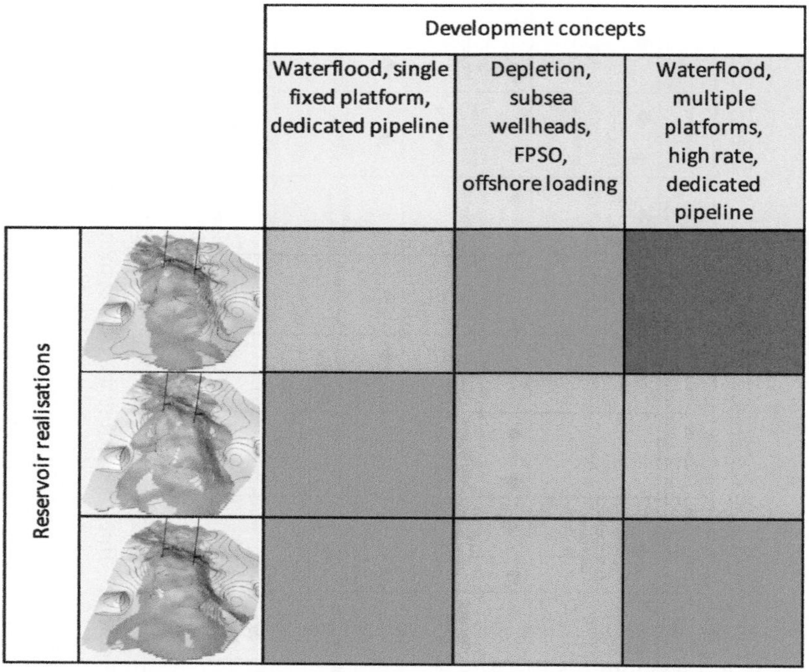

		Development concepts		
		Waterflood, single fixed platform, dedicated pipeline	Depletion, subsea wellheads, FPSO, offshore loading	Waterflood, multiple platforms, high rate, dedicated pipeline
Reservoir realisations				

Figure 8.8 Concept selection matrix using updated development concepts from threading.

The criteria for the traffic lighting are typically based on economic indicators, such as NPV, NPV/NPC or IRR. In this example, the single fixed platform (blue) appears to be the most robust. Building capacity with the large investment (light green) fails in the case of a low volume reservoir realisation. The FPSO (pink) is optimal only in the low reservoir volume realisation since the costs are low with no water flood and minimum facilities, but capacity constraint and lower recovery factor penalise this concept when reservoir volumes are higher. The criteria for traffic lighting can be extended beyond the simple cash flow derived indicators, and the number of development concepts and reservoir realisations need not be limited to three.

The extended test for each concept would use not just economic criteria, but other considerations, some of which are listed in Table 8.6. Each box in the matrix can be scored on a scale of 1—10 and each criterion can be weighted to provide an overall weighted score for each development concept.

Table 8.6 Extended test of development concepts using a scorecard approach.

Criterion	Weighting	Concept 1	Concept 2	Concept 3
	1—5	1—10	1—10	1—10
NPV	5	8	5	4
NPV/NPC	4	7	7	4
Payback time	3	7	5	5
Maximum exposure	3	7	9	5
Safety	5	6	5	7
Environment	5	6	7	4
Technical complexity	3	8	2	4
Optionality	3	4	3	7
Local content	2	5	3	6
Weighted total score		216	176	166

In this example, technical complexity scores high if the project is very achievable, and low if technically challenging. Optionality refers to the potential to enable other options, such as oversizing processing capacity to allow future tie-ins to the infrastructure. The weighted total score is the sum of the product of each score multiplied by its weighting. The Excel = SUMPRODUCT function is useful to calculate this.

This is a rigorous and auditable approach, which can capture the concerns of many stakeholders and incorporate aspects of the project beyond the technical and economic, including social and political considerations. It is not a linear process, and iterations around the development are to be expected. The outcome of the process should be a preferred development concept that allows the project to progress from the Select stage into the Define stage.

The process described is applicable to major capital projects, and probably unnecessary for decisions on smaller projects. It may be considered as over-prescriptive within some organisations, who prefer to make a selection of the development concept based on more instinctive logic.

8.6 Risk management of the development concept

Following the selection of a preferred development concept, and despite having tested against a set of investment criteria, a large amount of

246 Petroleum Economics and Risk Analysis

uncertainty remains as to the actual project outcome as it moves into the Define stage. This section will consider the risks associated with the uncertainties, how the risks can be identified and then managed.

8.6.1 Spider diagrams

A useful tool at any stage of a project is to run sensitivities on the input assumptions to investigate the impact on an outcome. The outcome may be measured in terms of any of the indicators introduced in Chapter 7. In this section, sensitivity analysis is used to investigate the impact of uncertainties on the NPV for a selected development concept, considered as the most important of the indicators. It will also be demonstrated that sensitivity analysis can be used to generate ideas for alternative concepts.

The tax and royalty example from Chapter 6 is briefly summarised in Table 8.7, but with the additional ability to vary selected inputs using a multiplier (MULT). To investigate the impact of a 30% increase in the oil price, the MULT for oil price is set to 1.3 and the resulting NPV(8) noted ($557 m). Input values are varied one at a time, to the extremes corresponding to the uncertainty range advised by the relevant authority.

Table 8.7 Summary of tax and royalty economic model with ability to vary assumptions.

Commercial and fiscal assumptions Tax & Royalty terms			MULT								
Oil price	50	$/bbl	1.30								
Oil price escalation	3	% p.a.	1.00								
General rate inflation	5	% p.a.	1.00								
Fixed opex	5	% cumulative capex	1.00								
Variable opex	5	$/bbl	1.00								
Royalty rate	10	%	1.00								
Tax rate	60	%	1.00								
Capital allowance	25	% p.a. straight line	1.00								
Capex multiplier			1.00								
Production multiplier			1.00								

YEAR			1	2	3	4	5	6	7	8	9	TOTAL
Flat RT Oil Price	65	$/bbl										
MOD oil Price			67.0	69.0	71.0	73.2	75.4	77.6	79.9	82.3	84.8	
ANNUAL PRODUCTION		MMbbl	0	0	10	20	20	15	10	5	1	80
GROSS REVENUE		$m			710.3	1463.2	1507.1	1164.2	799.4	411.7	84.8	6055.8
PLATFORM CAPEX		$m	160	250								410
FACILITIES CAPEX		$m		200	150							350
RT NET CASH FLOW		$m	−200.0	−500.0	112.4	464.7	491.2	383.8	174.9	76.2	4.3	1007.4
CUM. RT NET CASH FLOW		$m	−200.0	−700.0	−587.6	−122.9	368.3	752.0	926.9	1003.2	1007.4	

Indicator	Value	Unit	Definition
cumulative net cashflow	1433.6	$mln	undiscounted cumulative net cashflow
pay-as-you-go	3	yrs	net cashflow first positive
payback time	4.5	yrs	cumulative net CF returns to zero
maximum exposure	551.3	$mln	minimum of cumulative net cashflow
NPV(8)	557.3	$mln	net present value (mid-year discounting)
IRR	37.8	%	internal rate of return
RROR	31.2	%	real rate of return
UTC	22.1	$/bbl	capex plus opex per barrel
PV UTC	26.5	$/bbl	PV (capex plus opex) per PV barrel
NPV / capex	0.59	-	NPV per unit of capex
NPV / PV Capex	0.69	-	NPV per unit of PV capex
NPV / PV(Capex + Opex)	0.45	-	NPV per unit of PV (capex plus opex)

For example, at the Assess stage the facilities engineer may advise that capex uncertainty is +40% and −30% relative to the base case assumption, as discussed in Section 6.3. Changing the capex will also change fixed opex in the model, and it must be noted that the resulting NPV is a combination of the change in capex and fixed opex.

Production would typically be changed by calculating the percentage change of P90 and P10 volumes relative to the P50 base case, if using probabilistically derived volumes. If using a deterministic approach to the volumetric range then the percentage change of the high and low cases relative to the mid case is recommended. This provides a clear source of the change.

The variation in other inputs should be guided by the stage of the project and the relevant expert, but Table 6.2 provides a guide if working at the end of the Select stage, that is the beginning of the Define stage.

The number of inputs to vary may of course be extended, but once each variable has been individually changed the results can be plotted either on a tornado diagram such as in Fig. 5.17, or on a spider diagram as shown in Fig. 8.9. The spider diagram plots NPV(8) on the y-axis, assuming it to be most important economic indicator. However, if another indicator is preferred, the same procedure can be followed by noting the impact of each change on that target. The data for Fig. 8.9 were arranged as in Table 8.8 and the graph created using an X−Y scatter plot in Excel.

Table 8.8 Input data for the spider diagram.

Parameter	% change from base	NPV $ mln
Reserves	30%	523.5
	0	280.6
	-40%	-43.2
Capex	40%	69
	0	280.6
	-30%	439
Variable Opex	40%	235
	0	280.6
	-30%	314
Fixed Opex	40%	254
	0	280.6
	-40%	307
Time to Production	50%	170
	0	280.6
	-50%	339
Oil Price	30%	557
	0	280.6
	-50%	-180

In this example, the impact of time to production was investigated by simply copying and pasting the production figures a year earlier and then a year later. Since in the base case first oil was at the start of Year 3, this was interpreted as a change of −50% and +50%. No other changes were made to the capex profile when shifting the first oil date, which may be considered as unrealistic, but the purpose of the exercise is to indicate the relative impact of changes to base case input assumptions. If this is considered too crude, then rescheduling of capex could be incorporated.

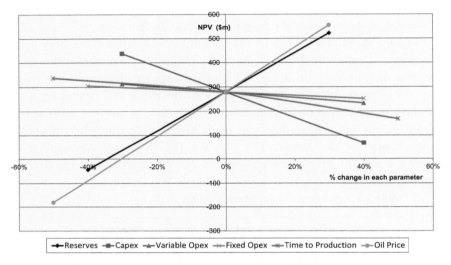

Figure 8.9 Spider diagram showing sensitivity of NPV(8) to variation in selected inputs.

The crossing point of all curves is the base case NPV, with no change in any input. The slope of the curves, which are rather linear in this example, indicate the sensitivity of NPV to each input. The extent of the curves indicates the impact of the expected range of uncertainty of each input on NPV. Both the sensitivity and the expected uncertainty range are important. For example, it is feasible to have an input to which NPV is very sensitive, but the input is not expected to vary widely, such as tax rate, in which case it is important to see that the forecast impact of realistic tax rate changes on NPV would be limited.

The outline joining all end points provides an overall impression of the robustness of the project to changes in the input. In the example, the downside of only two inputs takes the project below the break-even point of zero NPV, suggesting a robust project. In drawing these conclusions, it must be realised that each parameter has been changed one at a time, and a combination of changes could extend the perimeter of the spider web.

An optimist looking for upside in the project would start to focus on the inputs that create the high NPV outcomes, in this case from the upside in production and oil price, and from the lower end of the capex uncertainty range. Given that oil price is outside the control of the investor, the focus would fall on chasing the production upside and reducing capex. The pessimist looking at the potential downside would be concerned about the P90 production and the high side costs, as well as being aware that the low end oil price could yield a negative NPV.

While the economic model could be used to determine the change in any input which yields a break-even NPV (using, for example, the Goal Seek function in Excel), the spider diagram presents this immediately. In the example, it is clear that a 30% drop in oil price from the base case assumption yields zero NPV, while the project could withstand a capex increase of just over 50% before the project NPV is reduced to zero.

A final use of the spider diagram is to provide a guide to potential trade-offs in the development plan. Fig. 8.10 investigates the trade-off between spending incremental capex to achieve incremental reserves, both increments compared to the base case.

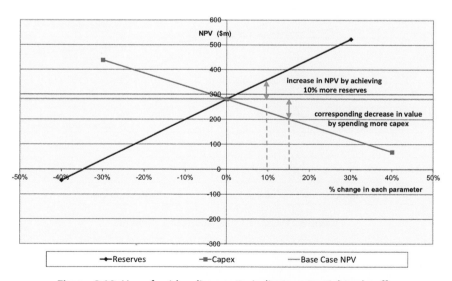

Figure 8.10 Use of spider diagram to indicate potential trade-offs.

Increasing reserves by 10% increases the NPV(8) by $80 m as shown by the blue arrow. This would be just offset by spending an additional 15% on capex as read from the graph, thus establishing the balance point of the trade-off in spending more capex to achieve more reserves. This is a crude analysis, but does provide some indication of the potential to optimise a selected development concept.

Using the same economic model and a spider diagram, Fig. 8.11 indicates a trade-off between capex and variable opex that is quite dramatic. It would be justifiable to increase variable opex by 55% in order to reduce capex by just 10%. This prompts the consideration of an alternative development plan that pays a tariff (in $ per barrel) for processing on a host platform, and reducing the project's platform development costs by removing processing equipment. The sensitivity analysis, presented on a spider diagram, is a somewhat crude approach but prompts ideas which can suggest alternative development concepts for the FDP. This investigation of plausible alternative development plans is exactly what is required at the Assess stage of a project.

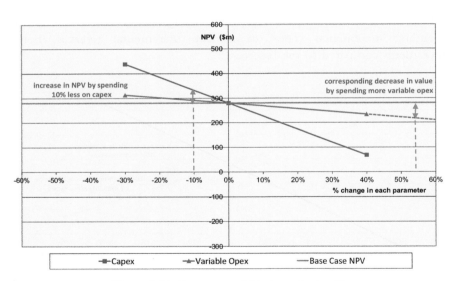

Figure 8.11 Capex vs variable opex trade-off.

In the example, capex costs are assumed to include both facilities (platforms, processing equipment, export pipelines) and well costs (drilling and completion, D&C). If capex was estimated separately for each of these and the NPV strongly impacted by D&C, this would prompt investigation into alternative D&C options, and could be used at the Assess stage of development planning.

In terms of managing the impact of the variables presented, some variables are within the control of the investor (capex, opex, project schedule) and others are outside of their influence (reservoir in-place volumes, oil price, tax levels). Variations from the base assumptions for these variables can present upside or downside in the project economic performance, as the spider diagram in Fig. 8.9 demonstrates. Whether the variable is within or outside the control of the operator, where significant upside or downside exists, effort can be made during development planning to manage the impact of these variations. The following section will discuss a tool that assists in managing the impact of so-called risk events.

8.6.2 Bow Tie model

The Bow Tie model has been used in industry for many decades as a method of identifying and managing risk. Its origin is often attributed to ICI (Imperial Chemical Industries), once a British chemical manufacturing giant, where the method was introduced in the late 1970s, and subsequently taken up and developed by many others, including Shell and BP in the oil industry (Ref. [6]).

The name of the model is clear from Fig. 8.12, the centre of the diagram being the knot in a gentleman's Bow Tie, but actually representing a **risk event**.

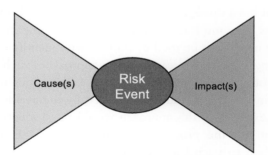

Figure 8.12 The Bow Tie model.

The term risk event requires some careful definition, but a very useful guide is to consider a risk event as something that happens to threaten the success of the project. Again, the term success needs to be qualified, and its interpretation will depend on the context in which risk analysis is being considered.

Table 8.9 suggests criteria for success in different contexts, representing different parts of the operator's organisation. It is valuable to have members of the departments agreeing to the criteria for success with respect to their own responsibilities, and recognising the success criteria for other departments and, importantly, for the overall project. Achievement of the criteria is often measured by a set of key performance indicators (KPIs).

Table 8.9 Examples of criteria for defining success of a project.

Department	Success measure
Subsurface	Delivery of reserves and production profile
Drilling	Well delivery within budget, on time, safely and with no lasting environmental impact
Operations and maintenance	Delivery of target uptimes Maintain production throughput target Remain within target for lost time incidents Operate with no lasting environmental impact
Commercial	Delivery of project value (NPV)
Contracts and legal	Agreement and implementation of contracts Compliance with contracts and legislation
Project management	Delivery of project value (NPV, cash flow) Delivery of project on time schedule

A risk event has a cause, or several causes, represented on the left-hand side of the Bow Tie, and leads to impact, or several impacts on the right-hand side. The model can be used by starting with identifying a risk event, and then tracing back to the causes using approaches such as root cause analysis or construction of a fault tree. Bell Laboratories introduced the latter method in the 1960s, and later adopted by the aviation and nuclear industries. An example of this approach would be to identify a risk event as an explosion on the process plant (which has HSE and economic impact), and then to trace back relevant causes of such an event. There would be a large number of contributing factors, including plant design, effectiveness of control and alarm systems and human activity.

An alternative approach to the left-hand side of the model is to start with a list of key uncertainties, such as the approach recommended in Section 8.2 where a long list of subsurface uncertainties was condensed down to a short list, representing key subsurface uncertainties. Each key uncertainty is then used to make a projection of a risk event. This follows the statement made in many texts that 'without uncertainty there is no risk', which is debatable, but a convenient link between risk and uncertainty in the context of the Bow Tie model.

For example, a key subsurface uncertainty may be that the depth of the hydrocarbon-water contact is poorly understood. A risk event linked to this would be that the volume of hydrocarbon in place may be significantly smaller than the base case assumption if the contact is shallow. Of course the contact could also be deeper giving rise to a significantly larger hydrocarbon in place volume, and in this case the risk event has a positive outcome. This is a reminder that risk events do not necessarily imply a loss or negative impact, though that is our instinctive interpretation of the word 'risk'. A positive outcome of the risk event would usually be called 'opportunity'.

Either working back from a risk event or working forward from key uncertainties is a valid approach to establishing the links between the risk event and the cause.

Moving on to the right-hand side of the model, a risk event such as a platform explosion or a low oil in place volume gives rise to an impact. A platform explosion could result in loss of production uptime, the requirement for a repair and injury to platform operators. A lower oil in place volume will almost certainly impact the project revenue when compared to the base case. With some effort, both such impacts can usually be measured in monetary terms (say, $). Loss or deferral of the revenue stream and cost of repair are readily converted into loss of NPV using the economic model. Sadly and somewhat uncomfortably, injury to personnel can be expressed in $ terms, as actuaries in insurance companies place a compensation value on personal injury.

At this point, it is appropriate to introduce a specific context for risk, quite different to that discussed in exploration risk in Chapter 4, being the Quantitative Risk Assessment (QRA) definition below.

QRA definition   RISK = PROBABILITY × IMPACT

In this context, probability refers to the chance of a risk event occurring, on a scale of 0%—100%, or 0.0—1.0. Impact refers to the cost (or benefit) of the outcome of the risk event, which can be measured in monetary terms ($). Applying these units it now becomes clear why this approach is called quantitative risk assessment, since risk is now expressed in dollars.

QRA definition (with units)

RISK ($) = PROBABILITY (0.0—1.0) × IMPACT ($)

This is how insurance companies set their premiums for a policy, by multiplying the impact of an accident with the probability of the accident occurring. In the context of a development project, we can do the same.

To manage the risk as defined above, there are two levers to pull. Either we take measures to reduce the probability of the risk event occurring in the first place, or we put in place measures to reduce the impact after the risk event has occurred. For example, to reduce the risk of an explosion, the operator could design gas detection system to shut down the process as soon as a significant gas leak occurs to reduce the probability of explosion, or construct blast walls around critical items of equipment to reduce the impact of the explosion, should it occur, or indeed both.

In this simple example, both actions were referred to as measures, but it is more useful to distinguish between those actions that reduce probability (called controls) and those that reduce impact (called contingencies), as shown in Fig. 8.13, for two main reasons. Firstly, the distinction allows the benefit of controls and contingencies to be measured independently, assisting in their justification, since each will come at a cost. Secondly, checks can be made by asking 'are the controls in place in the design of the equipment' and 'are the contingencies genuinely able to be implemented'. It can be a fatal mistake to hold the belief that a contingency is in place, only to find that when called upon, it cannot be completed.

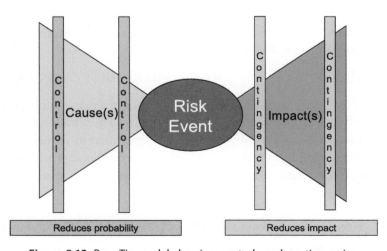

Figure 8.13 Bow Tie model showing controls and contingencies.

As a simple example, imagine planning a well to develop a thin reservoir which is expected to contain significant lateral shales within the reservoir section, and seismic control is poor, such that the depth of the top reservoir carries some uncertainty. The well is planned to be drilled horizontally through the reservoir and completed with a subsea wellhead, limiting the economic viability of later well intervention.

Each uncertainty is addressed individually as the cause of a risk event, and controls and contingencies are suggested to manage a risk event, as shown in the images in Table 8.10. The tables represent the Bow Tie image, but structured in a format that simplifies the recording of the elements.

Table 8.10 Bow Tie approach to manage risk events in well design.

Cause	Risk Event	Impact
Uncertainty on location of top reservoir due to poor seismic	Horizontal well sub-optimally placed in the reservoir (too shallow or too deep)	Early water breakthrough, with loss of oil reserves
Controls		**Contingencies**
Drill a pilot hole prior to drilling horizontal section	Example 1	Side-track and re-drill horizontal section to optimise depth within the reservoir

Cause	Risk Event	Impact
Lateral shale uncertainty could result in continuous shale baffles	Poor fluid connectivity within the reservoir	Reduced oil recovery
Controls		**Contingencies**
Drill a slant well through the reservoir rather than a horizontal well	Example 2	Hydraulic fracture of wells to improve connectivity

In Example 1, the proposed control to drill a pilot hole will add significantly to the well cost and somewhat extend the project schedule, and the Value of Information (VOI) method introduced in Chapter 5 could be used in its justification. It is an achievable technical control. The technical feasibility of the proposed contingency to re-drill the horizontal section as a side-track of the first wellbore would need to be checked. If the previous metal casing shoe was close to the reservoir section, then the side-track may need to be started from within a casing, which could be technically unfeasible, or at least slow and costly. For the contingency to be feasible, the casing design should be adapted to allow a side-track to be performed, otherwise the contingency is not realistic. This is an example of the need to design in advance to ensure that the planned contingency can be performed, if required.

In Example 2, the control is feasible and addresses the cause of the risk event, thus reducing the probability that the development suffers from poor reservoir connectivity, which would result in a loss of oil reserves.

The practicality and cost—benefit analysis of the contingency of hydraulic fracturing of the well should be investigated at the planning stage, since it would be a costly operation in a subsea well, and would rely on being able to mobilise equipment to carry it out. If the analysis proved this to be unfeasible, then the contingency recorded is unrealistic, and therefore not a true contingency.

The two risk events may also be linked, since both controls refer to the design of the well — the drilling of a pilot hole, and the trajectory of the well through the reservoir (horizontal well or slanted). Accepting one control may influence the option to put in place another, so that some compromise is required, or the risks need to be ranked. The use of the risk matrix in the next section will assist in determining the ranking of the risk events.

8.6.3 The risk matrix

The axes of the risk matrix represent the probability of a risk event occurring and the impact of that event, should it occur. Fig. 8.14 shows a typical risk matrix, in which the probability scale is in fractions from 0.0 to 1.0 and the impact axis is measured in dollars. It is colour coded in terms of severity of the risk, with a particular risk event plotted as a star.

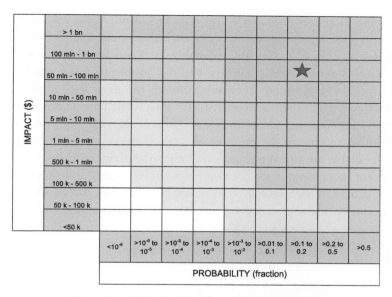

Figure 8.14 Risk Matrix with a risk event plotted.

The scales on the axes may be tailored to the specific type of risk and the risk tolerance of the company. For example, the probability axis may be represented in terms of frequency of an event, measured in number of incidents per year. This would be particularly relevant to safety incidents, measured in lost time incidents (LTIs) per annum or per unit of man-hours, with the probability estimate for a particular risk event being based on historical records. The impact axis can be scaled according to the particular project. In the example, the maximum value on the scale might represent total loss of the project at the point of maximum exposure. This may not even be the maximum on the scale — an incident such as the Macondo Deep Water Horizon disaster in 2010 would put the maximum impact beyond $50 billion.

The industry has developed many forms of the basic risk matrix shown. The impact in Fig. 8.14 is represented in dollars, but the impact may be measured in other terms such as impact on project cost, schedule, HSE or reputation. Arguably, all of these aspects can be converted into dollars, which is a convenient way of normalising the scale.

The colour coding on the risk matrix represents the degree of concern that the risk event represents, remembering that in QRA, risk is the product of impact and probability. In the pink zone, the risk would be unacceptable and such a risk event requires to be managed to move the risk out of this area. In the orange zone, acceptance of such risk would require consent at a high level within the company, probably at Board level. If not accepted then actions are required to move the risk profile into the green zone, which would be deemed acceptable, though the project manager would need to approve such acceptance. In the white zone, the risks are deemed insignificant and no further action is required.

In order to change the position of a risk event on the matrix, the two levers available are the application of controls to reduce the probability of the event occurring, and contingencies to reduce the impact of the event, should it occur. Fig. 8.15 illustrates the movement of the risk profile due to the application of controls and contingencies.

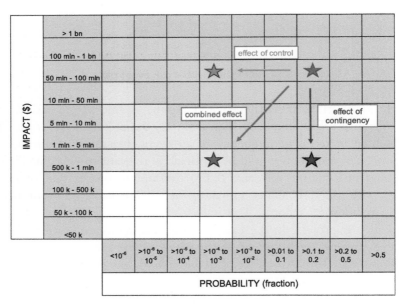

Figure 8.15 Effect of controls and contingencies on risk profile.

The combined effect is to reduce the risk, measured in dollars, and move the risk event into the acceptable zone. However, implementation of the controls and contingencies will have incurred a cost, and now a cost–benefit calculation can be performed to justify these measures. There will come a point at which the cost–benefit analysis suggests that no further measures can be justified on an economic basis. If the risk event were to remain in the orange zone after the measures, then a judgement will be required either to accept the risk or to apply further measures on a basis other than economics.

An example of this challenge is the decision of whether to flare excess produced gas or to invest in equipment to reinject gas. The reinjection costs may outweigh the benefit of reducing the risk of penalty payments, but may still be justified on the basis of following best operating practice or maintaining reputation. Despite the process of QRA, an element of judgement is often still required.

Having performed the analysis described, it is essential to record the results so that the project can continue to be well managed. A risk register is a valuable tool to capture the efforts in QRA and to provide the project manager with a useful tool to track project progress.

8.6.4 The risk register

The risk register records the analysis performed using the Bow Tie model, and adds the crucial action plan, along with the timing and party responsible for ensuring the actions are taken to implement the controls and prepare for the contingencies. Fig. 8.16 shows an example of a risk register using the reservoir risks introduced earlier, and using an identifier R to represent reservoir-related risks.

The final three columns may be included to indicate what the residual risk would be after performing these actions. However, unless these actions are actually performed, then the residual risk could be misinterpreted as being the current risk assessment, and some users prefer to omit the final columns, or to use a separate table to show the post-control and post-contingency residual risk.

The risk categories may be broken down into separate columns to reflect impact on project cost, project value, timing and HSE and colour coded rather than containing dollar values.

The risk register is a live document used by the project team to track the management of project risk. Each discipline will have a separate sheet detailing the relevant risks, and the document will be reviewed and updated periodically. As actions are completed, the risk profile will change. The project manager can use the tool to ensure that actions promised are carried out according to the schedule detailed in the register, which will become a key project management tool.

Identifier	Risk Cause	Risk event	Probability (%)	Risk Impact ($m)	Risk ($m)	Control	Contingency	Action Plan	Start/Finish	Owner	Residual Probability (%)	Residual Impact ($m)	Residual Risk ($m)
R1	Top reservoir depth uncertainty	Sub-optimal positioning of well	30	5	Water breakthrough and loss of reserves ($1.5m)	Drill pilot hole	Re-drill horizontal section	Run economics of drilling pilot hole. Check casing design allows side-track.	March 20xx / May 20xx	P. Borer (wells) J. Cash (economics)	10	5	$0.5m
R2	Baffling from lateral shales	Poor fluid connectivity	50	10	Reduced oil recovery ($5m)	Drill slant well	Hydraulic fracture of well	Check impact of slant wells on recovery. Investigate feasibility of fraccing	March 20xx / June 20xx	R. Sands (subsurface) M. Steel (technology)	30	5	$1.5m

Figure 8.16 Risk register with pre-and post-measure assessment of risk.

8.6.5 A financial adviser's view on risk

Before moving on from development planning decisions, this section will use another definition of risk, this time in the context of a financial adviser's approach to risk. This specific definition of risk will be applied to making an oil field development decision. The example will integrate the use of decision tree analysis and reiterate an important point regarding the interpretation of the term expected monetary value (EMV).

When considering the riskiness of investing in shares in a company, the historic record of the fluctuation in the share price over a chosen period of time can be used to define risk by comparing the standard deviation of the price to the mean price. Fig. 8.17 illustrates this and shows in red the growth in value of a guaranteed government bond with a fixed yield of 3% per annum, for which the variation around the mean is zero. Note that this has assumed conventional circumstances in which the yield on a bond is positive. During the highly uncertain times of Covid-19 in 2020, bond yields turned negative, a reflection of nervousness in the market.

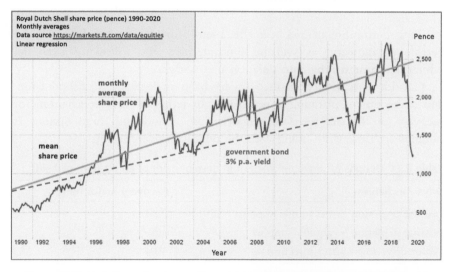

Figure 8.17 Variation of a share price compared to a typical government bond.

The green curve is the track record of the monthly average share price of Royal Dutch Shell (in pence) over a 30 year period, during which time the share price has increased, but there have been significant variations around this rolling mean, represented by the regression line in grey. The return on the investment in the shares exceeds that in the government bond, and traditionally this is to be expected. The investment in shares is more risky, and the investor should expect a higher return for taking that risk compared to the safer option of buying a government bond.

The impact on the share price of Covid-19 pandemic in 2020, combined with general over-supply of crude oil, is remarkable, reflecting an unprecedented rate of decline in oil price over the 30-year period.

The risk of investment opportunities can be measured by calculating the ratio of the standard deviation of the price divided by the mean. This ratio is known as the coefficient of variation (CV), or the relative standard deviation (RSD). It is a dimensionless term, which in the finance context represents the risk or volatility of the return compared to the amount of return expected, assuming that the future performance of the share will be similar to its past performance. Financial advisers are of course always at pains to stress that past performance is not a guarantee of the future.

$$\text{coefficient of variation (CV)} = \frac{\text{standard deviation}(\sigma)}{\text{mean}(\mu)}$$

This definition of risk can be useful for decision-making in the petroleum industry, as will be illustrated by the following example, which incorporates this into decision tree analysis.

Consider that an oil field discovery has been made in a new offshore basin, and development planning is underway. In terms of drive mechanism for recovery of the oil, reservoir engineers are considering three options as outlined in Table 8.11, along with the typical recovery factors.

Table 8.11 Options for drive mechanism in development of an offshore oil field.

Drive mechanism	Recovery factor (RF)
Natural depletion — no aquifer support	20%
Natural depletion — aquifer support	40%
Water drive from water injection	40%

The key uncertainty in this situation is the probability of the underlying aquifer being able to provide sufficient support to provide an effective water flood. To be able to achieve this, the aquifer needs to be large and permeable. In mature basins, this may become apparent after significant production from fields in the area, but in a new basin, knowledge of aquifer behaviour is very limited, since appraisal wells are usually targeted at the hydrocarbon bearing reservoir intervals, and, even if they penetrate the water column, very rarely test the dynamic performance of the aquifer. The aquifer properties usually only become apparent after several years of production from a new field.

With this key uncertainty in mind, Table 8.12 presents facilities development options under consideration, along with the implications for recovery. For simplicity, the example assumes that either the aquifer is active and supports an effective water drive giving 40% recovery, or is not active and the reservoir drive mechanism is depletion giving just 20% recovery. The range of uncertainty will be managed by varying the probability that the aquifer is active rather than creating a series of intermediate deterministic cases.

Table 8.12 Facilities options for development of an offshore oil field.

Facilities option	Implication	Recovery factor
1. Large platform with water injection from start	No dependence on aquifer performance	40%
2. Large platform with space for water injection to be retro-fitted	If aquifer is active, no requirement to retro-fit water injection	40%
	If aquifer not active, retro-fit water injection facilities, and drill wells, with slight delay in full recovery	40% with retro-fit but delay in production
	OR allow reservoir to deplete	20% with no retro-fit
3. Slim-line platform with no possibility for water injection	If aquifer active, no further facilities or wells expenditure, and achieve a full water flood recovery	40%
	If aquifer not active, build an additional platform dedicated to water injection	40% with additional platform but delay in production
	OR allow reservoir to deplete	20% with no additional platform

The key decision required is what facility to build, and the key uncertainty is the aquifer performance. Fig. 8.18 shows a decision tree structured to address this problem. Note the sequence of events; a decision on what facility to build is followed by the actual construction and some subsequent production, and only after a period of time does the aquifer performance become apparent, either strong or absent. In Option 1, the aquifer performance is not relevant, as the water injection (WI) from the start guarantees a water flood, and in fact the aquifer will never be given the opportunity to respond. Hence, there is no requirement for a chance node to represent the aquifer performance on this branch.

In completing the values on the tree, it is assumed that the best estimate of probability of a strong natural aquifer is 60%. There is of course uncertainty associated with this estimate, and subsequently a sensitivity analysis will be performed to see how this assumption affects the decision.

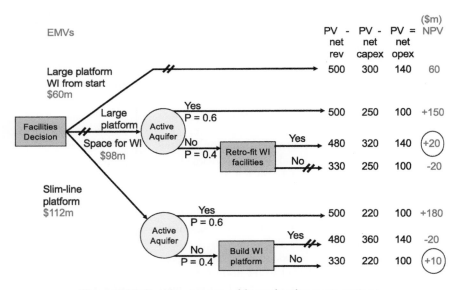

Figure 8.18 Decision tree to address development options.

The NPVs are calculated based on PV net revenue less PV capex less PV opex, each consistent for the development taking place on the branches of the tree. In rolling back from right to left and choosing the better NPV at decision nodes it is justified to add WI to the platform with space for WI in the case of no natural aquifer support. However, it is not justified to build a second platform in the slim-line development case if there is no natural aquifer support, since the cost of so doing creates a negative NPV, and it is better to simply allow the reservoir to deplete in this case.

NPVs are shown in green as the leaves or net payoffs at the end of the branches. Rolling back across the chance node representing the aquifer assumes a 60% probability of a strong (active) aquifer. The EMV of each decision at the root of the tree is shown in red. Based on the principles of decision tree analysis, the recommendation would be to build the slim-line platform since that option has the highest EMV. This problem is, however, worthy of further analysis.

Firstly, let us consider the impact of varying the probability of the natural aquifer providing a water flood. This can be readily performed on PrecisionTree using the sensitivity analysis option, or by hand, and the result is shown in Fig. 8.19.

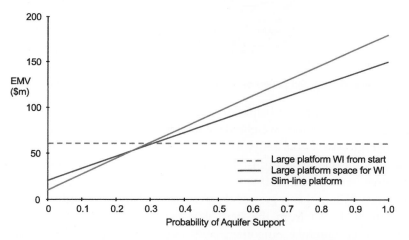

Figure 8.19 Sensitivity analysis on the assumption of aquifer performance.

This demonstrates that the EMV of selecting the heavy platform from WI from the start is not affected by the assumption of aquifer strength, but that the EMV of each of the other options is, as expected. Above approximately 30% probability of an active aquifer the better choice is the slim-line platform (Option 3), and below this the heavy platform with WI from the start (Option 1) is the optimum choice. Of key interest is the result that building the platform with space to retro-fit WI (Option 2) appears never to be the best choice. Perhaps then, the decision can be simplified to a choice between Option 1 and Option 3, as shown in Fig. 8.20.

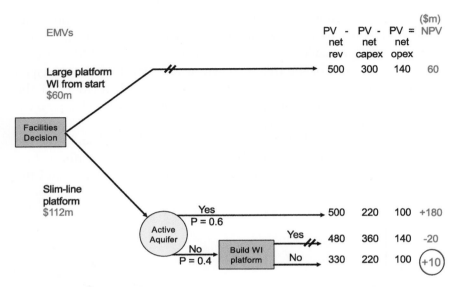

Figure 8.20 Simplified decision tree with two competing options.

This simplified tree is helpful to stress the significance of the term EMV, which is the probability-weighted value of future possible outcomes, and to consider the risk associated with the choices. The $112 m EMV for the slim-line option is the probability-weighted average of the two possible outcomes ($180 \times 0.6 + 10 \times 0.4$), essentially the mean (μ) value of the outcomes. While this is greater than the $60m EMV for building a heavy structure with WI from the start, it is the probability-weighted average of two very different outcomes, only one of which will be the actual outcome.

It is now useful to apply the definition of risk as used in the finance world.

$$\text{risk} = \text{coefficient of variation}(\text{CV}) = \frac{\text{standard deviation}(\sigma)}{\text{mean}(\mu)}$$

The standard deviation (s.d. in Table 8.13) can be calculated in Excel using the = STDEV.P function with input as a series of 60 values of 180 and 40 values of 10. For completeness, the risk for Option 2 is calculated, with results shown in Table 8.13.

Risk analysis and decision-making for development 267

Table 8.13 Risk and reward table for development options.

Case	Mean (μ)	s.d. (σ)	Risk (σ/μ)
1. Large platform with water injection from start	60	0	0
2. Large platform with space for water injection	98	64	0.65
3. Slim-line platform	112	83	0.74

The table demonstrates that there is a risk associated with each option. Option 1 delivers the lowest EMV but carries zero risk. Option 3 delivers the highest EMV, but also carries the highest risk, while Option 2 is intermediate level of risk and reward. By definition, EMV is the probability-weighted outcome, and in general should not be considered as a guarantee. Only for Option 1 could the EMV be considered as a guarantee since there is no chance event and thus no variation in the outcome.

It should now be clear that the choice between the options is not simply determined by the DTA rules of choosing the highest value at a decision node, since it may also be influenced by the degree of risk the investor is willing to take.

Fig. 8.21 presents the same results in a graphical manner, on a risk-reward plot. The reward is the EMV of the option, while the risk is the σ/μ of the actual outcomes.

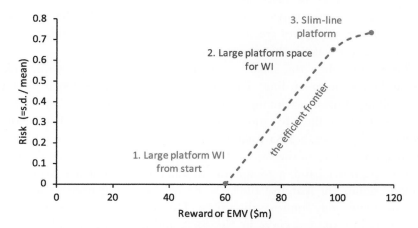

Figure 8.21 Risk-reward plot for development options.

This shows the risk implication of each option, but does not answer the question as to which is the best choice; high risk-high reward (Option 3), low risk-low reward (Option 1), or the intermediate Option 2. Economists would say that this is a matter of the investor's taste, or appetite for risk, which depends on the circumstances of the investor. This can be gauged by understanding the so-called utility value of each outcome, which will be developed in Section 8.6.6.

The curve on the graph also represents the efficient frontier, a concept that will be further explained in Chapter 10. In essence, one would ideally like to have an option that is placed on the extreme south-east corner of the plot, high-reward and low-risk. However, the closest that one can get to that, given the options available, is to be at one of the points on the curve — the curve is the 'efficient frontier'.

8.6.6 Utility

An individual's appetite for risk can be assessed by measuring the utility of a loss or gain. Utility is the satisfaction a person derives from the consumption of a good or service. In the case of our development decision example, the good is the EMV of the project. The individual may be a person or a company, and the utility associated with monetary values can take different shapes. A set of typical utility curves is shown in Fig. 8.22, differentiating between risk-averse, risk-neutral and risk-seeking individuals.

If the individual is risk-neutral then the utility curve will be the straight line, also known as the EMV-line, and replacing EMV with its equivalent utility will always make the option with the highest EMV the preferred choice. When the stakes are small, most individuals will be risk-neutral. In this case, the stake is the investment required.

However, when the stakes are high, most individuals become risk-averse, being the green line. The marginal utility of incremental gains is diminishing. Imagine winning a lotto prize of $5 million. This would provide huge utility to most individuals. If by freak chance that individual won another $5 million, the utility would not be doubled — the marginal utility of the second tranche of winnings is less than the first.

On the other hand, a $5 million windfall for a billionaire would not provide significant marginal utility — he or she would need a far larger sum to significantly improve the utility. The billionaire's utility would be represented by the red curve in Fig. 8.22, and he or she would display a risk-seeking attitude.

Many small- to medium-sized E&P companies will exhibit risk-averse behaviour, but larger organisations may be more risk-seeking. A single discovery for Shell has a small marginal utility. It takes a much larger event such as the acquisition of BG in 2016 to make a significant difference to its business. This acquisition is reported to have cost $53 billion.

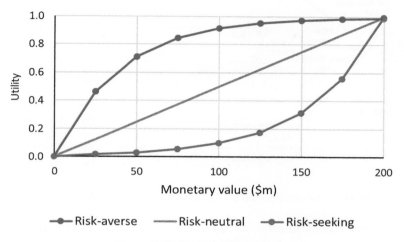

Figure 8.22 Typical utility curves.

The development decision is now performed from the perspective of two different companies, one being risk-averse (NewCo) and the other risk-taking (MagCo), but now replacing the NPVs of the outcomes with their corresponding utilities shown in Fig. 8.22. Using the normal DTA method, when the utilities are rolled back using the respective probabilities, the Expected Utility (EU) is calculated.

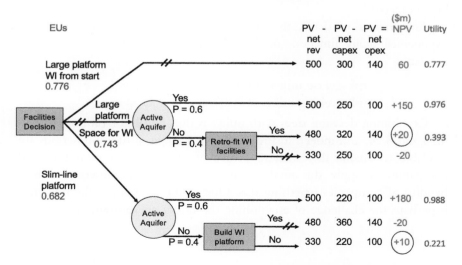

Figure 8.23 Development decision for risk-averse NewCo using Expected Utility (EU).

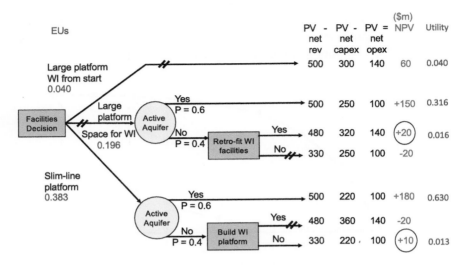

Figure 8.24 Development decision for risk-seeking MagCo using Expected Utility (EU).

As one would intuitively expect, based on utility, the risk-averse company favours the lowest risk option, the large platform with WI from the start, despite it having had the lowest EMV. The risk-seeking company is guided by the EU to the highest risk option.

If a company were risk-neutral then the option that maximises the EU would be the same as that which maximises the EMV. In this case, there is no need to apply a utility function, and the highest EMV would always be the appropriate choice. As mentioned, when the stakes are considered to be small, individuals tend to be risk-neutral, and utility does not need to be considered.

As a matter of detail, the curves in Fig. 8.22 were generated using exponential utility functions of the following forms, where b and R are constants:

risk-averse utility (U) of value x $\quad U(x) = 1 - e^{(-x/R)}$
risk-seeking utility (U) of value x $\quad U(x) = ae^{bx}$

Combining decision trees with utility is thus possible, as shown in Figs 8.23 and 8.24, and indeed in PrecisionTree a range of utility functions are available to enable this analysis. However, it is not common for E&P companies to apply this method, and the author has observed only a handful of companies practising this. One of the reasons for the reticence is perhaps the challenge of establishing the utility curve itself.

Many economics texts describe the utility curve as being somewhat subjective, and note that the curve will change with a company's circumstances such as access to investment capital, ability to repay debt and exposure to the vagaries of the oil and gas price.

It is also noteworthy that utility curves can extend to the negative outcome, as illustrated in Fig. 8.25, which rebrands the utility axis as pleasure for a positive outcome (gain), and pain for a negative outcome (loss). Close to the origin, win or lose $10, one is risk-neutral and feels relatively indifferent, but the pleasure of winning $20 is less than the pain or disappointment of losing $20.

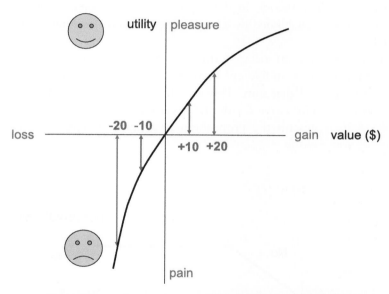

Figure 8.25 Utility curve extending to negative outcome.

On the gain part of the curve, the size of the investment is most likely increasing as the curve extends, and it is intuitive that one would feel more pleasure from winning $20 from a stake of $100 than say winning $50 from a stake of $1000.

On the loss part of the curve, most individuals and company senior managers will have an intuitive estimate of how much they could afford to lose in an investment before threatening their livelihood or business. If the exposure threatens this, then the investment should probably be avoided, though there are no fixed rules that can set the limits of risk that should be taken — it is again a matter of the individual circumstances. Making a series of losses that exceed the funds available and put one out of business is known as gambler's ruin, and there are many examples of investors who have fallen into that situation.

The real challenge in basing decisions on EU rather than EMV is establishing the utility curve. While the curve can be based on the historical decisions made by the investor, accepting that circumstances change over time, for those seeking a more quantitative method of establishing a utility curve, Mian (Ref. [7]) describes in detail how utility curves can be created by using preference theory. In this approach, the risk tolerance (R) of the decision maker is established by asking a question and repeating the level of the stakes.

The R-value is in monetary terms ($) and is the sum of money at which the decision maker is indifferent between a 50:50 chance of winning that sum and losing half that sum. This was the R-value used in the equation to construct a utility curve for the risk-averse decision-maker above.

Fig. 8.26 represents the proposition in a decision tree. Note that all values are **net** payoffs.

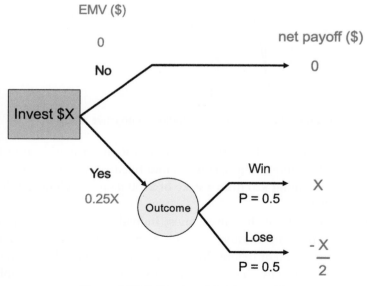

Figure 8.26 Estimating risk tolerance (R).

When asking this question of my university students the R-value is usually around $10. Losing $10 probably means sacrificing dinner. The R-value for my oil company employees on a training course is usually around $100, and for more established professionals $1000. Occasionally a risk-seeking individual will push to $10,000 on the assumption that he or she could withstand several losses in a row and still continue to play sufficient number of times to realise the EMV, which is of course positive. Utility is a matter of personal circumstance, and oil company's R-values will be in the millions of dollars, but vary widely depending on their scale.

The most important point to be made in this section is that EMV is not a guarantee. Only if the investor is risk-neutral should the EMV be used to decide which option to choose. When considering the riskiness of each option the variability of the outcome should be presented, and the definition of risk as the CV is useful. If one chooses not to calculate this, then at least one should point out the variation of the outcome on the decision tree. The probability chart generated in PrecisionTree, as shown in Fig. 8.27, is a further way to illustrate this. This alerts the decision-maker to the variability in possible outcomes and avoids disappointment when the actual outcome does not equal the EMV.

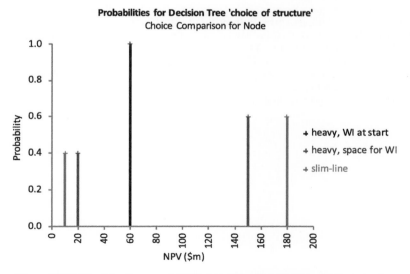

Figure 8.27 PrecisionTree probability chart for the development example.

The probability chart demonstrates that the heavy platform with WI from the start has 100% probability of delivering $60 m while the other options carry a much wider range of variability in the outcome. Combined with a presentation of the EMVs, this supports the concept of the risk involved in the options. Whatever method is used, whenever there is uncertainty involved in the outcomes, it is responsible to emphasise that EMV should not be assumed to be a guarantee for any single development decision.

If the investor were to develop 200 fields, exactly the same as the example, then the average outcome of developing with the slim-line platform would of course be $112 m. In the long-run, this is the best decision. However, the development decision for each field is only taken once, and the investor should be aware of the financial adviser's definition of risk, consider the risk-reward balance and the company's utility curve. This is a more structured approach than making decisions based on instinct, and is more important as the stakes get high.

References

[1] P. Ringrose, M. Bentley, Reservoir Model Design — A Practitioner's Guide, Springer, 2015, ISBN 978-94-007-5496-6.
[2] D. Kahneman, Thinking, Fast and Slow, Penguin, 2011, ISBN 978-0-141-03357-0.
[3] R.L. Plackett, J.P. Burman, The design of optimum multifactorial experiments, Biometrika 33 (4) (June 1946) 305—325.
[4] M. Jamshidnezhad, Experimental Design in Petroleum Reservoir Studies, Gulf Professional Publishing, 2015, ISBN 978-0-12-803070-7.
[5] F. Jahn, M. Cook, M. Graham, Hydrocarbon Exploration and Production, second ed., Elsevier, 2008, ISBN 978-0- 444-53236-7.
[6] CGE, BowTieXP, The Next Generation BowTie Methodology Tool, 2015. https://www.icao.int/safety/SafetyManagement/SMI/Documents/BowTieXP%20Methodology%20Manual%20v15.pdf.
[7] M.A. Mian, Project Economics and Decision Analysis, PennWell, 2002, ISBN 0-87814-855-8.

CHAPTER 9

Incremental projects and decommissioning

Contents

9.1 Identifying incremental projects	275
9.2 Incremental project economics	281
9.3 Ranking incremental project opportunities	287
9.4 Decommissioning	295
9.4.1 The economic limit and cessation of production	295
9.4.2 Decommissioning activity	299
9.4.3 Decommissioning costs	301
9.4.4 Fiscal treatment of decommissioning costs	302
9.4.5 Decommissioning cost liabilities	304
References	306

9.1 Identifying incremental projects

The preceding chapters have focused on bringing the oil or gas field from an exploration prospect to first production, but as Fig. 1.1 illustrated, the majority of a field lifecycle is in the production phase. A brown field generally refers to a field in production and a mature field to one in the production decline phase. These terms tend be applied loosely, but in both situations the main capital investment in the field development has been made and a base case production and net cash flow profile have been established.

In addition to maintaining the base production and net cash flow, the operator will seek incremental project opportunities to improve the project performance. Such activities usually require incremental capex or opex investment to deliver benefit through improved future production or reduced overall future costs. As Fig. 9.1 illustrates, the operator will seek incremental project opportunities throughout the producing lifetime of the field, even before the planned first production start date. The production profile in red represents the planned activities, known as the base case or reference case profile. This may also be called the no further activity (NFA) case. The grey lines represent potential improvements to the profile through incremental project activities, and Q_{ab} is the production rate at the time of abandonment. The reasons for reaching the decision to abandon the field are discussed in Section 9.4.1. The word decommissioning has generally replaced the term abandonment, which has implications of simply leaving equipment in place, which is not the case.

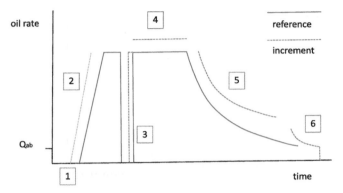

Figure 9.1 Incremental project opportunities through the producing lifetime.

The objective of many incremental project activities is to improve the point-forward project economics, and examples are described in Table 9.1. In every case, the effect of the incremental project is compared to the reference case, which is a fundamental principle of evaluating incremental projects.

Table 9.1 Typical incremental project activities.

Item	Objective	Examples of activity
1	Advance first oil date	Early production system (EPS) using tie-in to an FPSO or converted semi-submersible rig Temporary tie-back of wells to existing infrastructure Pre-drill development wells prior to commissioning of main production facility
2	Accelerate build-up to plateau	Prioritise drilling sequence to focus on high-productivity wells first Improve drilling and completion practices using upgraded technology and equipment Hire additional drilling rigs or add a second drilling derrick to the platform
3	Reduce planned shut-down time	Increase manpower during planned shut-down Invest in higher specification equipment to extend run-life and reduce maintenance

Table 9.1 Typical incremental project activities.—cont'd

Item	Objective	Examples of activity
4	Increase plateau rate	De-bottleneck processing plant Invest in additional production facilities
5	Slow down decline rate	Drill infill wells Workover wells to improve flow rate Apply enhanced oil recovery (EOR) methods
6	Extend the economic lifetime	Invest in technology to reduce operating costs Apply EOR methods Adapt facilities to act as a processing hub for third party production and share opex

In each of these examples, additional expenditure is committed in the expectation of improved project performance. In economic terms, the improvement can be measured by evaluating the net cash flow of the incremental project, which is addressed in Section 9.2.

Some incremental projects are justified to achieve compliance with regulation or with performance standards, rather than on pure economic grounds. For example, changes in legislation controlling allowable concentrations of oil-in-water in produced water discharge to sea would require an incremental project either to perform produced water reinjection (PWRI) or to apply additional treatment prior to discharge. The plant upgrade would maintain the licence to operate but not yield an incremental revenue stream, and the economic analysis would focus on the most cost-efficient method of achieving the target performance rather than on maximising value.

In terms of production profiles, incremental projects fall into two categories: acceleration projects and additional recovery projects. Items 1 to 4 in Table 9.1 will generally be acceleration projects, meaning that they enable the reserves to be produced earlier than in the reference case, but do not add to reserves. The area under the production vs. time curve to the point of abandonment remains the same. With the exception of the workover opportunity, the examples in Items 5 to 6 in Table 9.1 are aimed at adding recovery of hydrocarbons to the field. Section 9.2 will address the economic analysis of both types of incremental project.

The opportunity to increase the production throughput constraint (Item 4) by de-bottlenecking existing facilities is both challenging to identify and to evaluate. Fig. 9.2 illustrates selected equipment that should be familiar from the Concept Selection process described in Section 8.3: reservoir—wells—facilities—export. With the maximum design rate for each of the items in the process shown, it is a simple matter to identify that the bottleneck constraining production is the oil export pump.

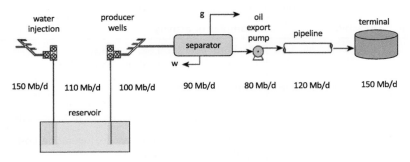

Figure 9.2 Identifying bottlenecks — the choke model.

There is evidently an opportunity for an incremental project to de-bottleneck the production by replacing the oil export pump with a larger capacity item, say a 100 Mb/d pump. This would incur immediate capex for the replacement and additional future opex. The incremental benefit, however, would only be 10 Mb/d, since the next bottleneck in the system becomes apparent — the separator, with its maximum design rate of 90 Mb/d. One by one, opportunities for de-bottlenecking will continue to appear, each representing a potential incremental project.

The single incremental project suggested is the replacement of the oil export pump. As presented, the single pump is actually a critical item in the process, meaning that if it is out of service, the whole process will shut down. To shut down the process, replace the pump and run the process back up to full operating conditions may take several days to weeks, depending on the complexity of the replacement and the plant. This will significantly reduce the uptime of the system, uptime being the actual producing time divided by the total time. On an annual basis an offshore oil platform operator would aim for around 95% uptime, excluding any planned shut-down, also known as turn-around (TAR). An improvement on the suggestion to replace the pump would be to install a second pump in parallel and possibly avoid shut-down, or to implement the pump replacement during the next planned TAR.

On the assumption that the separator capacity could be increased beyond 90 Mb/d by adjusting operating pressure or weir height, and that the well productivity could be improved by perhaps changing tubing size or tubing head pressure, installing pump capacity greater than 100 Mb/d could be considered. A chain of potential incremental project opportunities may be evaluated, so that the decision on the new pump capacity is future-proofed for subsequent activities. If identified in advance, as in the example, several improvements may be lumped together in the incremental project proposal.

The example is based on one parameter, the oil rate. However, during the field life the operating parameters will change, so while in the early stages of production the oil handling facilities are candidates for de-bottlenecking, as water production inevitably increases the water handling plant is likely to become a constraint. If the gas:oil ratio (GOR) climbs, the gas handling plant may become a constraint, or if the reservoir pressure depletes, the water injection plant may become the production constraint.

When the rates are clearly laid out, identifying the incremental production benefit of the de-bottlenecking is trivial. In practice, however, creating the equivalent of Fig. 9.2 is challenging, since it requires the collation of information from many parts of the operator's organisation: subsurface, wells, process engineering, operations, pipeline operator and terminal managers. There is limited unified commercial software available to integrate flow performance through all parts of this system. Despite the challenges, many operators persist to link the elements together to investigate creating value through de-bottlenecking opportunities, and name the approach 'choke-modelling'. A functioning choke model, continually updated with current operating data such as rates, fluid composition and pressures, can identify the constraint at any particular time and be used to evaluate potential incremental de-bottlenecking projects throughout the producing lifetime of the field.

De-bottlenecking part of the production system is an example of an acceleration project, in which oil is produced early compared to the reference case, but no reserves are added. The associated production profile is illustrated in Fig. 9.3A.

The examples for Item 6 in Table 9.1 are incremental projects that extend the economic lifetime of the project. Recall that the economic lifetime is determined by the date at which the economic limit (EL) has been reached; the gross revenues are equal to the costs, so the net cash flow is zero, and at this point the production rate from the field is noted in Fig. 9.1 as the abandonment rate, Q_{ab}. As an example, drilling an additional well in an undeveloped extension to the field as an incremental project would deliver three benefits, as illustrated in Fig. 9.3B.

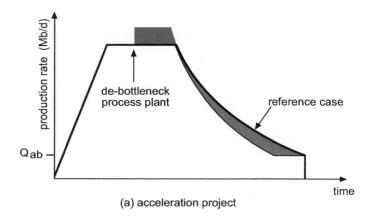
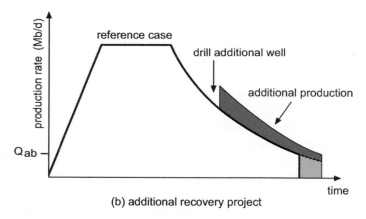

Figure 9.3 Acceleration versus additional recovery incremental projects.

The benefits of the additional recovery project, which can be balanced against the incremental costs, are
- the additional well adds reserves that would not have been produced in the reference case (dark green)
- the economic lifetime is extended so that the original part of the field can produce for a longer time before the EL is reached (light green)
- the abandonment activities can be deferred, reducing the PV cost of abandonment.

The multiple benefits of additional recovery projects often make them more attractive than pure acceleration projects, which essentially rely on the effect of discounting the net cash flow to deliver a positive NPV.

9.2 Incremental project economics

In performing the economic analysis of incremental projects, it is essential to identify all costs and revenues that are attributable to the incremental activity and to exclude those that will occur regardless, or those not affected by the decision to implement the project. To make the differentiation, the reference case must be clearly established, since an underestimate of the reference case may make the incremental project appear to have more value than the marginal benefit it truly generates. The reference case may also be known as the base case or NFA case.

As Fig. 9.3B demonstrates, there may be more elements to the evaluation of the incremental project than purely the incremental cost and incremental revenue from production, as in this case the deferral of abandonment cost plays a role. To ensure that all aspects of the incremental project are captured in the analysis, the most reliable approach to the evaluation is to run two economic models. One model is the reference case and the other is a composite case, being the reference case plus the incremental project. The difference between the two models is then attributable to the incremental project. This approach is sometimes known as with/without economic analysis and can be generalised as

$$\text{composite} - \text{reference} = \text{incremental}$$

The alternative method is to isolate all incremental costs and benefits and run the DCF analysis on the incremental net cash flow, as shown in Fig. 9.4.

In both methods, a new reference date is set for the future net cash flow, being the point of first investment in the incremental project.

The recommended method is to run two cases, composite and reference, and to compare the difference in net cash flow. As long as a robust model for the reference case is already available, this should not require any re-building of the model to capture the complexities of taxation, tariffs or opex. The following example demonstrates the method, assuming the drilling of an additional well in an undeveloped extension to the field during the decline phase, as shown in Fig. 9.5. The reference case shows declining production over a 9-year period, with the contribution from the additional well coming in the third year of the profile shown. In evaluating the incremental project, the reference date is set at the point of expenditure on the additional well — the original project reference date is irrelevant as we are considering only point-forward net cash flows.

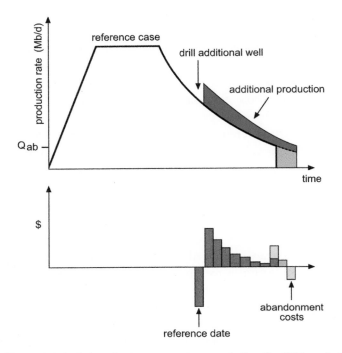

Figure 9.4 Isolating the incremental net cash flow for DCF analysis.

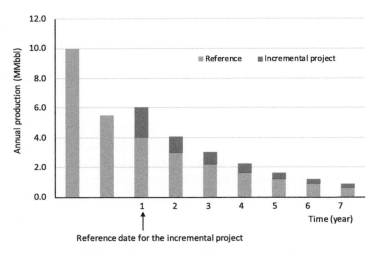

Figure 9.5 Incremental production from an infill well.

Using the economic model for a tax and royalty system introduced in Section 6.5.1, the reference case is analysed in Table 9.2. For clarity, the model selected does not include inflation.

Table 9.2 Reference case cash flow model.

Commercial and fiscal assumptions			Tax and Royalty terms								
Oil price	50	$/bbl									
Inflation	0	% p.a.									
Fixed opex	5	% cumulative capex									
Variable opex	5	$/bbl									
Royalty rate	10	%									
Tax rate	60	%									
Capital allowance	25	% p.a. straight line									

INCREMENTAL PROJECT YEAR				1	2	3	4	5	6	7	TOTAL
Oil Price	50	$/bbl									
REFERENCE PRODUCTION		MMbbl		10.0	5.5	4.1	3.0	2.2	1.7	1.2	0.9	0.7	12.2
INCREMENTAL PRODUCTION		MMbbl									
GROSS REVENUE		$m		500.0	274.4	203.3	150.6	111.6	82.6	61.2	45.4	33.6	1383.7
REFERENCE CASE CAPEX		$m									0
INCREMENTAL CAPEX		$m									0
TOTAL CAPEX		$m	0	0	0	0					0
Cumulative Capex			800	800	800	800	800	800	800	800	
Capital allowance - REFERENCE		$m									0
Capital allowance - INCREMENTAL		$m									0
TOTAL CAPITAL ALLOWANCE		$m									0
Unit Variable opex	5	$/bbl									
VARIABLE OPEX		$m	50.0	27.4	20.3	15.1	11.2	8.3	6.1	4.5	3.4	138.4
FIXED OPEX		$m	40.0	40.0	40.0	40.0	40.0	40.0	40.0	40.0	40.0	280.0
TOTAL OPEX		$m	90.0	67.4	60.3	55.1	51.2	48.3	46.1	44.5	43.4	418.4
Royalty Rate	10	%									
ROYALTY		$m	100.0	54.9	20.3	15.1	11.2	8.3	6.1	4.5	3.4	215.8
TAX DEDUCTIONS		$m	190.0	122.3	80.7	70.1	62.3	56.5	52.2	49.1	46.7	634.2
TAXABLE INCOME		$m	310.0	152.1	122.6	80.5	49.3	26.1	9.0	-3.7	-13.1	749.5
Tax Rate	60	%									
TAX		$m	217.0	106.5	73.6	48.3	29.6	15.7	5.4	-2.2	-7.9	495.9
GROSS REVENUE		$m	500.0	274.4	203.3	150.6	111.6	82.6	61.2	45.4	33.6	1383.7
CAPEX + OPEX (TECHNICAL COST)		$m	90.0	67.4	60.3	55.1	51.2	48.3	46.1	44.5	43.4	418.4
GOVERNMENT TAKE		$m	317.0	161.3	93.9	63.3	40.7	23.9	11.5	2.3	-4.5	711.7
DECOMMISSIONING COST									100.0			
NET CASH FLOW		$m	93.0	45.6	49.1	32.2	19.7	10.4	-96.4	-1.5	-5.2	153.6
CUMULATIVE NET CASH FLOW		$m	93.0	138.6	187.7	219.9	239.6	250.0	153.6	152.1	146.9	
					NPV(10)		34.9	$m			

In this reference case, the project net cash flow becomes negative in Year 6, noting that the reference date is set at the time at which it is intended to drill the additional well. The project should not continue in Year 6, and a decommissioning cost has been included in Year 5, highlighted in red. As previously, cash flow items are shaded in yellow. The NPV(10) of the remaining project cash flow in Years 1–5 (highlighted by the outline) is calculated as $34.9 m, using mid-year discounting. Note that the NPV is significantly influenced by the decommissioning costs, but is still positive. If the owner were trying to divest this project, this would be a reasonable valuation.

The same model but now run as a composite case is shown in Table 9.3, including the drilling of an additional well, at a cost of $20 m, and this well delivers additional production commencing in the same year as it is drilled.

Table 9.3 Composite case cash flow model, including additional well.

Commercial and fiscal assumptions		Tax and Royalty terms									
Oil price	50	$/bbl									
Inflation	0	% p.a.									
Fixed opex	5	% cumulative capex									
Variable opex	5	$/bbl									
Royalty rate	10	%									
Tax rate	60	%									
Capital allowance	25	% p.a. straight line									

INCREMENTAL PROJECT YEAR				1	2	3	4	5	6	7	TOTAL
Oil Price	50	$/bbl									
REFERENCE PRODUCTION		MMbbl	10.0	5.5	4.1	3.0	2.2	1.7	1.2	0.9	0.7	13.1
INCREMENTAL PRODUCTION		MMbbl			2.0	1.1	0.8	0.6	0.4	0.3	0.2	5.3
GROSS REVENUE		$m	500.0	274.4	303.3	205.5	152.2	112.8	83.5	61.9	45.8	1693.6
REFERENCE CASE CAPEX		$m										0
INCREMENTAL CAPEX		$m			20							20
TOTAL CAPEX		$m	0	0	20	0						20
Cumulative Capex			800	800	820	820	820	820	820	820		
Capital allowance - REFERENCE		$m										0
Capital allowance - INCREMENTAL		$m			5	5	5	5				20
TOTAL CAPITAL ALLOWANCE		$m			5	5	5	5				20
Unit Variable opex	5	$/bbl										
VARIABLE OPEX		$m	50.0	27.4	30.3	20.5	15.2	11.3	8.4	6.2	4.6	169.4
FIXED OPEX		$m	40.0	40.0	41.0	41.0	41.0	41.0	41.0	41.0	41.0	326.0
TOTAL OPEX		$m	90.0	67.4	71.3	61.5	56.2	52.3	49.4	47.2	45.6	495.4
Royalty Rate	10	%										
ROYALTY		$m	100.0	54.9	30.3	20.5	15.2	11.3	8.4	6.2	4.6	246.8
TAX DEDUCTIONS		$m	190.0	122.3	106.7	87.1	76.4	68.6	57.7	53.4	50.2	762.2
TAXABLE INCOME		$m	310.0	152.1	196.6	118.4	75.8	44.2	25.8	8.5	-4.3	931.4
Tax Rate	60	%										
TAX		$m	217.0	106.5	118.0	71.0	45.5	26.5	15.5	5.1	-2.6	605.1
GROSS REVENUE		$m	500.0	274.4	303.3	205.5	152.2	112.8	83.5	61.9	45.8	1693.6
CAPEX + OPEX (TECHNICAL COST)		$m	90.0	67.4	91.3	61.5	56.2	52.3	49.4	47.2	45.6	515.4
GOVERNMENT TAKE		$m	317.0	161.3	148.3	91.6	60.7	37.8	23.9	11.3	2.0	851.9
DECOMMISSIONING COST										100.0		
NET CASH FLOW		$m	93.0	45.6	63.7	52.4	35.3	22.7	10.3	-96.6	-1.7	226.4
CUMULATIVE NET CASH FLOW		$m	93.0	138.6	202.3	254.6	289.9	312.6	323.0	226.4	224.6	
					NPV(10)		99.7	$min				

The incremental production, capex and capital allowances are shown in red. The fixed and variable opex and tax calculations reflect these additions. The project should not continue in Year 7, and a decommissioning cost has been included in Year 6, highlighted in red. The decommissioning date has now been deferred by a year, as one of the benefits of the incremental project. The NPV(10) of the remaining project cash flow in Years 1—6 is calculated as $99.7 m, using mid-year discounting.

Therefore applying the principle that

$$\text{composite} - \text{reference} = \text{incremental}$$

the value added by the incremental project is

$$\$99.7 \text{ m} - \$34.9 \text{ m} = \$64.8 \text{ m}.$$

The same result can be reached by isolating the incremental cash flow elements, and this is shown in Table 9.4, using the same commercial and fiscal assumptions.

Table 9.4 Analysis of the purely incremental cash flow elements.

INCREMENTAL PROJECT YEAR					1	2	3	4	5	6	7	TOTAL
Oil Price	50	$/bbl										
REFERENCE PRODUCTION		MMbbl	0.0	0.0	0.0	0.0	0.0	0.0	0.0	0.9	0.0	0.9
INCREMENTAL PRODUCTION		MMbbl			2.0	1.1	0.8	0.6	0.4	0.3	0.2	5.3
GROSS REVENUE		$m	0.0	0.0	100.0	54.9	40.7	30.1	22.3	61.9	12.2	309.9
REFERENCE CASE CAPEX		$m										0
INCREMENTAL CAPEX		$m			20							20
TOTAL CAPEX		$m	0	0	20	0						20
Cumulative Capex			800	800	820	820	820	820	820	820	820	
Capital allowance - REFERENCE		$m										0
Capital allowance - INCREMENTAL		$m			5	5	5	5				20
TOTAL CAPITAL ALLOWANCE		$m			5	5	5	5	0			20
Unit Variable opex	5	$/bbl										
VARIABLE OPEX		$m	0.0	0.0	10.0	5.5	4.1	3.0	2.2	6.2	1.2	31.0
FIXED OPEX		$m	0.0	0.0	1.0	1.0	1.0	1.0	1.0	41.0	1.0	46.0
TOTAL OPEX		$m	0.0	0.0	11.0	6.5	5.1	4.0	3.2	47.2	2.2	77.0
Royalty Rate	10	%										
ROYALTY		$m	0.0	0.0	10.0	5.5	4.1	3.0	2.2	6.2	1.2	31.0
TAX DEDUCTIONS		$m	0.0	0.0	26.0	17.0	14.1	12.0	5.5	53.4	3.4	128.0
TAXABLE INCOME		$m	0.0	0.0	74.0	37.9	26.5	18.1	16.9	8.5	8.8	181.9
Tax Rate	60	%										
TAX		$m	0.0	0.0	44.4	22.7	15.9	10.9	10.1	5.1	5.3	109.1
GROSS REVENUE		$m	0.0	0.0	100.0	54.9	40.7	30.1	22.3	61.9	12.2	309.9
CAPEX + OPEX (TECHNICAL COST)		$m	0.0	0.0	31.0	6.5	5.1	4.0	3.2	47.2	2.2	97.0
GOVERNMENT TAKE		$m	0.0	0.0	54.4	28.2	20.0	13.9	12.3	11.3	6.5	140.1
DECOMMISSIONING COST										-100	100.0	
NET CASH FLOW		$m	0.0	0.0	14.6	20.2	15.6	12.2	106.7	-96.6	3.5	72.8
CUMULATIVE NET CASH FLOW		$m	0.0	0.0	14.6	34.8	50.4	62.6	169.4	72.8	76.3	
					NPV(10)		64.8	$m				

Indicators			definition
Cumulative net cash flow	169.4	$m	Undiscounted cumulative net cash flow
Pay-as-you-go	-	yrs	Net cash flow first positive
Payback time	-	yrs	Cumulative net CF return to zero
Maximum exposure	-	$m	Minimum of cumulative net cash flow
NPV(10)	64.8	$m	Net present value (discounting at 10%)
IRR	#NUM!		Internal rate of return
UTC	15.7	$/bbl	Capex plus opex per barrel
PV UTC	14.6	$/bbl	PV (capex plus opex) per PV barrel
NPV/Capex	3.24	-	NPV per unit of capex
NPV/PV Capex	3.56	-	NPV per unit of PV capex
NPV/PV(Capex+Opex)	1.54	-	NPV per unit of PV (capex plus opex)

This confirms the result of using the recommended method of running a reference case and a composite case, but much care is required in creating a purely incremental case. For example, in Table 9.4 it is first necessary to calculate when the project net cash flow becomes negative with and without the incremental project in order to know how far decommissioning is deferred, and how much additional production from the reference case is gained, along with the associated opex. However, once the incremental cash flow items are isolated, it allows further economic indicators to be calculated.

The most comprehensive method of evaluating the incremental project is to run the reference case, create the composite case and then extract the difference in each element of the cash flow, which is how Table 9.4 was created.

Because the net cash flow in the year of expenditure on the project is positive, and the time periods are in years, it is not possible to calculate some of the basic indicators such as pay-as-you-go, payback time and maximum exposure. Low-cost incremental projects such as infill wells, workovers and efficiency upgrades often achieve payback within a year, so if these indicators are required, the cash flow analysis would need to be performed on a monthly basis.

Note also that the IRR cannot be calculated, since in this case there is no negative net cash flow in Year 1 of the incremental project. Even if that were the case, there would be multiple changes of sign of the net cash flow, noting the negative net cash flow at the end of the project, and hence multiple possible solutions to the IRR calculation. This is often the case with incremental projects, and caution is advised in the use of IRR.

The efficiency indicators in this example are very positive, recalling that for a major capital project, investors would hope for NPV/PVCapex to exceed 0.4 and NPV/PV(Capex + Opex) to exceed 0.2. This incremental project is very attractive. The only drawback is that compared to a major capital project the reward is low. It is an example of a low-risk, low-reward project, but if the operator can find many such projects in a mature field, the sum of the relatively small rewards can make very profitable business.

In terms of the indicators from the incremental project, the only one that is truly additive is the NPV, meaning that the NPV of the increment is the NPV of the composite less that of the reference case. None of the other indicators are additive in this way, and isolating the incremental net cash flow is the only way of determining the efficiency indicators IRR, UTC and NPV/PVCapex.

It is worth noting two points about the reference case. Firstly, if this is underestimated, then the incremental project may look too attractive — more value becomes attributed to the incremental project than it deserves. If it is over-estimated, then the incremental project may never get the time to deliver its reward. For example, if the production facilities prematurely fail to meet the technical standard that allows continued licence to operate, then the facility may be shut in before the full benefit of the incremental project is realised.

Secondly, the reference case becomes the platform for launching incremental projects, which can make such projects economically very efficient. Even when the reference case is running close to the EL, the facilities may be used as a hub to attract business from other fields by processing and exporting their production. In mature basins such as the North Sea, there are many examples of field life being considerably extended by taking this approach, sharing operating cost and earning tariff payments for processing services. This allows the original field to produce for longer, add reserves and defer decommissioning costs. Some operators refer to safeguarding the reference case as 'managing the base', recognising that this not only delivers the reference case value but provides the opportunity to implement valuable incremental projects.

9.3 Ranking incremental project opportunities

Building on the reference case activities as a foundation, many potential incremental projects can be identified throughout the producing lifetime of the field. As for green field development, the company is usually capital or net cash flow constrained, making it necessary to prioritise the incremental project opportunities.

The economic screening and ranking process described in Chapter 7 can be applied, checking which projects beat the hurdle rate of return and ranking based on NPV, or using a capital efficiency ratio such as NPV/PVCapex. The hurdle rate of return may be lower than for major capital projects if the risk associated with the incremental project is considered to be low. Due to the relatively low capex required for many incremental projects, the payback time is often short and the IRR high. The Excel = IRR function will compute the IRR as infinite if the payback is less than a year and the economic analysis is based on annual net cash flows. The RT net cash flow is discounted using the cost of capital, or else an inflationary element should be built in to the discount rate when discounting an MOD net cash flow, as described in Section 6.7.

However, economic indicators are not the only consideration in the deciding the priority of incremental projects. An agreed hierarchy of criteria is useful to rank the overall attractiveness of projects and the order in which they should be performed, or determining which projects should be deferred or rejected.

While the relative importance of ranking criteria may vary between operating regions and with time as constraints on the business change,

Table 9.5 Suggested criteria for ranking incremental projects.

Ranking criterion	Comment and examples
HSE	Usually stated as the primary consideration to justify projects that protect health, safety and the environment.
Licence to operate	Projects that secure an operator's licence to operate are often seen to be a fundamental requirement and require immediate action.
Value	Measured in $, commonly as NPV, and balances both the revenue and the cost of a project. Reflects a primary economic interest of the shareholders in terms of delivery of value.
Efficiency	Measured by any or all of the efficiency indicators, IRR, NPV/PVCapex, unit technical cost, the indicators are useful to maximize a return on a constrained resource such as investment capital. Can also extend to maximizing the return on a throughput capacity, e.g. $/bbl/d.
Reserves	Addition of reserves may be an important annual target to satisfy previous promises to shareholders or host governments, and analysts may be looking carefully at this performance indicator.
Production rate	Similar to reserves addition targets, annual production rates may be an expectation of shareholders or governments. Failure to meet contractually agreed gas sales volumes may invoke significant penalties.
Cost	Where cash flow is a significant constraint, project cost (either capex or opex, or both) becomes an important indicator, but may be balanced by short project payback time or low UTC, and so this crude criterion should be considered alongside the capital efficiency ratios.
Complexity	Technically or commercially challenging projects may reduce their attractiveness. A cross plot of complexity vs a value or efficiency indicator (NPV or NPV/PVCapex) is used to visualise the trade-off between the reward and the likely difficulty of executing the incremental project.
Risk management actions	Using a tool such as the Bow Tie model (Section 8.6.2) the effort and cost required to manage the risk event to an acceptable residual risk are considered. Projects with costly or complex risk management actions will be demoted in the ranking. As with complexity, a cross plot against the reward will visualise this criterion.

Table 9.5 is a suggested set of ranking criteria. They are listed in what many operators would consider to be the approximate order of their relative importance, though this is very much a matter of taste and circumstance for any individual company. Where a long list of criteria is used, each may be weighted to provide some overall score.

It would be unusual for the first two criteria in Table 9.5 not to appear close to the top of the list. Most operators would state that a primary intention is to perform with no harm to people and no lasting negative impact on the environment. Incremental projects that ensure that this intent remains feasible are often approved without hesitation.

Without continued licence to operate, as granted by the host government or its representative authority, the operating company may not be allowed to continue to pursue its business. Incremental projects that secure this are of course of high importance. It would a rare decision not to preserve the licence to operate, but if, for example, the cost of making a facility safe is high, the solution may indeed be to cease production and start decommissioning.

A common front-end statement from the CEO in the annual report is that the company is in business to create value on behalf of the shareholders. On this basis, value is a powerful decision criterion, and it conveniently captures revenue, cost and, if using NPV, timing and cost of capital for the incremental project.

Qualitative matrices such as the one shown in Fig. 9.6 can be useful in screening out unattractive incremental projects, in this case using the project NPV as the criterion for reward and the project complexity as the risk. Complexity is assessed qualitatively. A routine workover on a

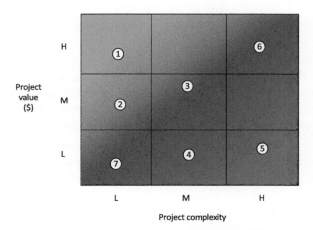

Figure 9.6 Qualitative risk reward matrix of project complexity vs value.

platform well with successful precedents would be low complexity, but installing a sub-sea multiphase booster pump in deep water would be a complex project.

Proposals that fall outside of the acceptable envelope (Project 5) may be deferred, considered for re-design to reduce complexity, or simply rejected. Shelving projects for later consideration is not recommended, as this risks wasting time recycling the evaluation. Projects 1—3 may be deemed acceptable, while Project 6 is a candidate for reviewing the risk management actions or adapting the project design.

With the usual abundance of incremental project opportunities put forward, it is possible keep pursuing opportunities such as Project 7 above, without necessarily adding value. BP developed an analysis for determining when to stop adding incremental projects, trademarked and released to the industry as WETS, a wellwork evaluation tracking system. The method was developed as part of the well management process within the company to identify the cost per barrel of wellwork, a form of incremental project. Details can be found in Refs. [1,2].

Fig. 9.7 is an adaptation of the plot used in the process. The x-axis represents cumulative incremental production from the series of wellwork projects, while the y-axis is the cumulative cost of the jobs. The points on the curve are not plotted in chronological order, but rather in order of increasing cost per barrel. The projects near the origin are those that produce incremental oil at the lowest cost per barrel, while those at the end of the curve are destroying value as they actually lose production, for example, adding perforations in a well but those perforations actually start to produce water and reduce the lift performance of the well, thus producing less oil.

Due to the shape of the curve, this form of plot has become known as a scorpion plot, with the sting in the tail. It is a useful representation of the actual performance of historic incremental projects as it can help to distinguish the type of project that delivers production at low cost, problems from the type of project that carries a high cost per barrel and threats from the grouping of projects that can destroy value. If there is a target or hurdle cost per barrel set by the company then the type of project that exceeds this cost per barrel should be rejected or reworked before approval. The cost per barrel is the gradient of the curve.

In analysing the plot, one is looking for trends in the grouping of projects moving from the successes to the failures, and learning from this to identify and plan future wellwork. The technique may be applied to any form of incremental project.

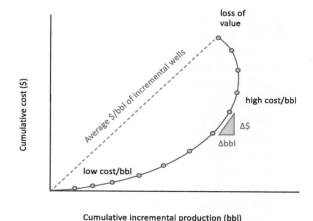

Figure 9.7 Cost per cumulative incremental barrel after WETS.

In the scorpion plot above, no detailed economics has been performed using the DCF technique and the composite less reference case analysis, and yet the approach is instructive. It is not always necessary to run a full economic analysis to identify and approve incremental projects, particularly if they are of the type that needs to be performed to retain the licence to operate. Pre-tax cost per barrel analysis may suffice.

However, when the stakes become higher, it is recommended to take the composite minus reference case net cash flow approach to ensure that all the implications of the proposed incremental project have been taken into account. There may be less obvious effects of the project on taxation, contracts, third party constraints and timing of decommissioning. For example, if the incremental project takes production levels above a threshold, the tax position of the project may change — some PSCs have a cost oil or royalty level which depend on production, and some tax and royalty systems have a tax rate that is a function of production rate or even of cumulative production.

It is recommended that when the investment levels are considered to be high, the simple pre-tax approaches are used for screening incremental project proposals, but full composite and reference case DCF models are used for decision-making. In most cases, the DCF model developed for major capital project decisions can also be used for the incremental project analysis, so it is not necessary to build the incremental project economic model from scratch.

Just as for major capital projects, it is also important to consider the uncertainties and risks associated with an incremental project, as the following example will demonstrate.

Wells in a mature oil field development, producing at a plateau production rate, have started to produce some water. Simultaneously a calcium carbonate scale has started to form around the safety-critical subsurface safety valve (SSV). The operator is removing the scale by shutting in each well annually for 10 days to retrieve and descale the SSV, and acidising the tubing to remove scale deposits. This defers production from each well to the time at which total production drops from the plateau rate and there is capacity in the process system to produce the deferred oil. The total deferral from the wells is 10 Mb/d, and this will occur for the next 3 years, after which all the deferred oil is expected to be produced in a single year. The net-of-tax value of each barrel is $4. All costs will be calculated net-of-tax allowances.

The incremental project opportunity suggested by a production chemist is to add a scale inhibitor chemical to the already installed chemical injection line, which is currently injecting a corrosion inhibitor. This will require extra chemical costs and the installation of an additional chemical storage tank, but will reduce the cost of the annual SSV retrieval and tubing acidisation, and also prevent any further deferral of oil. Fig. 9.8 shows the well completion.

Fig. 9.9 shows the incremental net cash flow for this project, indicating that the proposal is essentially an acceleration project. No additional reserves are produced. The PV benefit of the accelerated production in Years 1–3 is offset by the PV loss of that same total production in Year 4 since in the reference case this is when the deferred oil would have been produced.

Mid-year discounting the net cash flow at a 10% cost of capital yields NPV(10) = $110,000. This appears attractive and the project is approved. The storage tank is installed, and the chemical is injected into the first well. Unfortunately the well stops flowing, and the subsequent investigation determines the problem to be incompatibility of the commingled injection chemicals with the oil and water from the reservoir when mixed at the bottom hole conditions of temperature and pressure. There is a plug of gunk (a sticky residue) at the bottom of the well. This will require an intervention into the well and a rethink of the incremental project.

What could have been done to test the robustness of this project from an economic perspective? The NPV(10) of $110,000 looked attractive as a

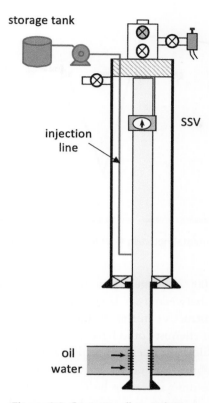

Figure 9.8 Current well completion.

stand-alone figure. This is the balance of PV revenues (above the line in Fig. 9.9) and the PV costs below the line. These are

$$\text{NPV} = \text{PV revenue} - \text{PV costs}$$
$$\text{NPV} = \$710,000 - \$600,000 = \$110,000$$

A simple sensitivity analysis on the assumptions shows that an overrun of \$110,000 in cost (18% cost overrun) or a \$110,000 underperformance in revenue (15% reduction in revenue) would cause the project to simply break-even with an NPV of zero. Given the uncertainties discussed in Chapter 6, these are modest variations, probably well within a realistic range of uncertainty in the assumptions. The incremental project is very finely balanced, and this should be understood prior to approval.

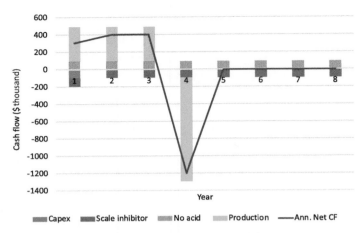

Figure 9.9 Incremental net cash flow for scale inhibitor injection project.

While the $110,000 is a large sum in most individuals' view, in this case it is the difference between two large numbers. The sensitivity analysis would have been instructive in this example.

If the production chemist could have anticipated failure, then a decision tree risk analysis could have been performed as shown in Fig. 9.10. In this example, the probability of success is carried as an unknown (p). The probability of failure must therefore be $(1 - p)$. The prize associated with success is $110,000. A regret cost must be associated with failure and in this example it is assumed that an intervention cost for the well is $300,000. The mode of failure and method of remediation would need to have been foreseen to arrive at this assessment. Failure to remediate would mean losing the well, with an associated regret of $800,000.

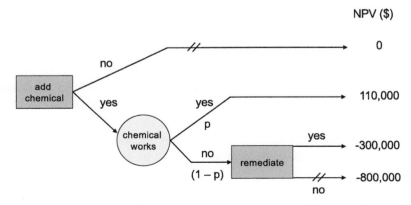

Figure 9.10 Decision tree used to calculate break-even probability of success.

The NFA option is the reference case with and EMV of zero. The probability of success (p) required for the project to be equivalent to NFA is calculated by

$$110,000 \times (p) + (-300,000) \times (1-p) = 0$$
$$410,000 \times p = 300,000$$
$$p = 73\%$$

This probability should have been used prior to approving the project to assess the chemist's opinion of success, as they would need to be at least this confident that the chemical would work for the project to demonstrate a positive NPV.

When asked for a probability of an event, individuals often struggle to express an opinion. Setting a benchmark probability of success can be a useful prompt to elicit a thoughtful opinion. In the example DTA has been used without an input probability — the approach has backed out the break-even requirement.

This is an actual field example. The chemist had tested the commingled chemicals in the laboratory, mixed with produced oil and water, but not at the downhole conditions of temperature and pressure, where they became insoluble. Fortunately, the chemical dosing was applied only to one well before the incompatibility became apparent. The formulation of the injection chemicals was later adapted and the project was successful in solving the scaling problem for all wells.

Incremental projects can add significant value to the asset and be the lifeblood during the late-life production. Eventually all fields will run out of opportunities and face the inevitable decommissioning activities.

9.4 Decommissioning

As mentioned in Section 9.1, decommissioning is the commonly used term, replacing abandonment, which has connotations of not acting to remove facilities and make safe wells, pipelines and any other equipment that is left in situ. This section will no longer use the term abandonment.

9.4.1 The economic limit and cessation of production

The time at which decommissioning is required is at the Cessation of Production (CoP) date, which can come about either because the field has reached its economic limit (EL) or its technical limit (TL).

Prior to granting the operator the approval to stop production, the government's regulatory authority will need to be convinced that all reasonable measures have been taken to prolong field life and to maximise the economic recovery. For example, in the United Kingdom, the Oil and Gas Authority (OGA) requires several years' prior notice of the intended CoP, along with a plan for decommissioning. Guidelines are published for this process (Ref. [3]).

The TL is caused by technical limitations that cause production to stop, such as irreparable equipment failure, or equipment failing to meet integrity inspections by working outside of design parameters.

The EL can be determined by running an economic limit test (ELT), based on the cash flow analysis, and in the simplest case is when the operating costs equal the revenues, but all evaluated on a pre-tax basis.

The earliest limitation, EL or TL will determine the CoP date. Neither the TL nor the EL is the same as licence expiry date, as the field will usually remain economic beyond this and the licence may be extended or operations taken over by another entity such as the NOC of the host country.

The EL is normally when operating costs equal the net revenues (net of royalty and any property-specific overhead charges), noting that the operating costs should be restricted to those costs that will be eliminated if the project stops. Fig. 9.11 illustrates the EL occurring at the intersection of costs of production and revenues, and shows that reducing costs, increasing revenues or preferably both can defer the EL.

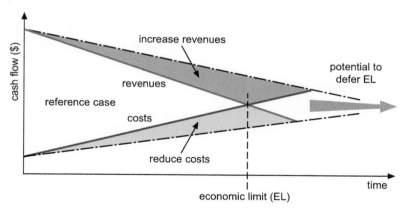

Figure 9.11 The economic limit (EL) and potential deferral of decommissioning.

The first lever the operator usually applies to extend the EL is to reduce operating costs. This can be done through increasing efficiency, standardising equipment and procedures, reducing the supplier costs, shortening the supply chain to reduce equipment storage costs, sharing facilities and supply chains with other assets or neighbouring operators, extending periods between maintenance and reducing staffing levels. Unfortunately, the lever that is often quickest to pull is the last in this series, and the industry has a reputation for hiring and firing in a cyclical manner. Manpower cost can represent a third of total operating cost and becomes a clear target for dramatic and immediate cost reduction.

Following the oil price crash in 2014, companies reduced operating costs dramatically. Rystad Energy, a consultancy group, reported a 35% reduction in maintenance and operations service prices in the period 2014–16 (Ref. [4]). With the arrival of a record drop in oil price in 2020, it became even more challenging to create similar operating cost savings after the previous streamlining.

The other lever that Fig. 9.11 highlights is increasing revenues. The incremental projects discussed in this chapter become opportunities to do this, but each project has to be screened and ranked and meet the company's investment hurdles, while allowing it to operate with sufficient net cash flow to survive. If a significant capital investment is required, such as implementing an EOR scheme, then the individual project net cash flow may become negative for a significant period before reaching pay-as-you-go time, and this may need to be carried by positive contributions from other projects to the corporate net cash flow. This stresses the importance of considering such incremental projects well before the EL approaches.

Revenues may be more quickly increased from tariffs by offering the production facility as a host to third party production. This has been the lifeline for many UKCS facilities, such as the Andrew Field developed by BP which became the production hub for tie-back of both the operator's and third party oil fields. This has extended the timing of the decommissioning of the facility beyond the CoP date of the stand-alone Andrew field and allowed production to continue in 2020 at just 3000 bbl/d.

Short payback incremental projects with low capex and small incremental opex are particularly useful to effectively increase production and hence revenues. Examples include well workovers to replace pumps, additional perforations in existing wells to access bypassed oil, well stimulations, reduction of separator inlet pressures or compressor upgrades for gas fields and upgrades of processing equipment to improve performance.

A third lever, that can extend the EL in a fiscal system that includes royalty payments, is to persuade the host government to reduce or abolish royalty on the field. Recall that royalty is based on gross revenues and is a pre-tax levy. At the margin, when costs equal revenues, there will be no taxable income, and therefore no tax, so the EL is reached when

$$\text{costs} = \text{revenues}$$
$$\text{opex} + \text{royalty} = \text{revenues}$$

Removing the royalty from the above balance creates a positive net pre-tax cash flow (i.e. taxable income), which is taxable. This is a win-win situation for the operator and the host government, as the government collects tax rather than collecting nothing, and the operator retains the net-of-tax cash flow and defers CoP, and hence decommissioning costs. Reserves are added, which support the value of the oil company and are in the interest of the host country in maximizing use of national resources.

Following lobbying by a collective group of operators in the UKCS, the UK Government abolished royalty in 2003, providing a lifeline to some 30 pre-1982 mature assets that had been paying royalty up to that point. As a point of detail, those fields gaining approval for development after 1982 were never subject to royalty payments.

Of course there may be short periods during the producing life of the field when opex exceeds the revenues and pre-tax net cash flow is negative, and yet the operator continues production. This may occur when investing in a second phase of development or a significant EOR project, but a positive net cash flow is forecast to return. Even without such capex investments, the net cash flow can become negative due to low oil or gas price. If, despite the operator's efforts to reduce costs, the cash flow remains negative, the operator has further difficult choices to make.

One common option is to continue production in the hope that product prices rebound. This avoids shutting in production, after which it can be slow to reinstate normal operating conditions, and retains staff, which avoids later recruitment. As long as the cash flow losses are sustainable, this defers the need to start decommissioning the equipment. Oil price in particular is highly volatile. In the price crash in early 2020, the \$20/bbl Brent crude low rebounded to \$40 within two months. Most offshore installations continued production through this period.

Even with longer term losses, the operator may choose to continue production in order to defer the decommissioning costs. If the opportunity cost of the capital required for decommissioning is greater than the daily negative net cash flow, the company may choose to keep the cash in the bank and earn interest while accepting the losses from the producing asset, rather than spend the capital on decommissioning. From a reserves reporting perspective it should be noted that production that is not economic cannot be reported as reserves, according to the Petroleum Resources Management Guideline (PRMS, Ref. [5]). It needs to be classed as contingent resources. This prevents analysts, who base their opinion partially on reserves, from overvaluing the asset, as such production is not adding value.

9.4.2 Decommissioning activity

The activities required to perform decommissioning are outlined in Table 9.6. This has been extracted from the UK's guidelines for decommissioning offshore oil and gas installations and pipelines (Ref. [6]). Such guidelines are typically issued by the host government authority, so that the operator can submit its specific plan for decommissioning a field, which is reviewed by the authority and revised by the operator prior to final approval to commence decommissioning.

Table 9.6 Decommissioning activities.

Item	Range of actions
Consideration of re-use	Alternative use for the depleted reservoir, e.g. carbon capture and storage (CCS) or enhanced oil recovery (EOR). Potential re-use or preservation of export pipelines.
Wells	Permanent isolation of the reservoir and all intermediate zones with the potential to flow, to provide sufficient barriers to flow and pressure. Removal of wellheads and conductors (for dry wells). Removal of sub-sea wellheads, manifolds and templates.
Production facilities	Under OSPAR Decision 98/3: Topside installations returned to shore for safe disposal. Steel jackets less than 10,000 tonnes to be removed for re-use, recycling and final disposal on land. Derogation of above guide for large steel jacket footings and concrete jackets. Removal of floating production facilities and sub-sea facilities to land, or re-use.

Continued

Table 9.6 Decommissioning activities.—cont'd

Item	Range of actions
Pipelines	Primary consideration for re-use. Not covered by OSPAR Decision 98/3, but aim to achieve a clear seabed, usually inferred to mean removal or burial of in-field and export pipelines. If not removed then left such as not to create hazard to marine environment, fishing, and consider corrosion impact. Clearance of debris either side of pipelines.
Site remediation	Debris clearance for a minimum radius around the installation. Verification of clearance with side-scan sonar. Remediation of land or seabed sites.
Waste	Hierarchical waste management plan with considerations of re-use, recycling or disposal.
Environment	Environmental appraisal to assess impact of the decommissioning plan on the environment including energy balance and impact of removal techniques such as explosives.
Cost and timing	Cost estimate to demonstrate cost-effectiveness. Schedule of activities.
Post-decommissioning	Monitoring of site. Management plan for infrastructure left in situ.

The specific actions required for these items will be determined by guidelines and legislation within the host country and by some international obligations for offshore installations. These include the 1992 Convention on the Protection of the Marine Environment in North East Atlantic ('the OSPAR Convention'), the United Nations Convention on the Law of the Sea of 1982 (UNCLoS) and the International Maritime Organisation (IMO) Guidelines and Standards. OSPAR Decision 98/3 included a ban on disposal of offshore installations at sea, and after coming into force in 1999, this has been adopted by some governments (including the United Kingdom).

Guidance and regulations are revised as decommissioning technology and experience develops. The United Kingdom is relatively advanced in decommissioning technology, guidance and legislation. Compliance with the legislation is assured by the Offshore Petroleum Regulator for Environment and Decommissioning (OPRED) which reviews decommissioning programmes submitted by the operator. On the United Kingdom Continental Shelf (UKCS) by 2020 there were over 600 offshore surface installations, over 3000 subsurface installations and some 3000 pipelines (Ref. [7]). Offshore decommissioning first started in the UKCS on the Argyll Field in 1992, and by 2020 some 9% of the platforms that had been installed had been decommissioned. In 2020, annual expenditure on decommissioning was around $2 bn and forecast to be $20 bn over the following decade (Ref. [8]).

However, the UK Oil and Gas Authority (OGA) report in 2019 (Ref. [9]) stated that their most recent estimate of the total cost of decommissioning the remaining offshore installations had reduced by some 17% between the 2017 estimate and that of 2019, from £60 bn ($75 bn) to £49 bn ($61 bn) in a like-for-like comparison. This reflected improved scoping of required work for well plug and abandonment (P&A), better execution practices for subsea wells, reducing rig rates, improved methods of platform and infrastructure removals and reduced platform running costs after CoP. Further potential reductions can be expected as operators and service companies learn from experience. This experience can be used to benefit regions outside the United Kingdom, and it is an ambition of UK companies to export this expertise.

The other major region building up experience in this technology is the US Gulf of Mexico (GOM). IHS Markit (Ref. [10]) reports that the GOM has more than 5000 oil and gas structures remaining in place, with 4000 already decommissioned. Many of these are in shallow water (10s of metres) compared to the UK North Sea (ca.100–200 m). Southeast Asia is expected to be the next focus of new decommissioning activity as fields become mature.

It is of interest to note that the technology exists to raise the platform topsides in one lift from the jacket and transported by barge onshore for dismantling, recycle and disposal. Specialist vessels such as the Allseas *Pioneering Spirit* have been constructed to float under the platform topsides by straddling the jacket, lift the topsides and transport the structure to shore (Ref. [11]). This avoids the costly need for a crane barge to lift topsides, which often weigh up to 25,000 tonnes. The Equinor-operated Statfjord A platform topsides weighs an extraordinary 48,000 tonnes, and it is planned to remove this in a single lift by upgrading the *Pioneering Spirit* catamaran vessel.

9.4.3 Decommissioning costs

As experience of decommissioning expands and the operational database builds up, cost estimation becomes more certain. Commercial software such as QUE$TOR, introduced in Section 6.3, contains a regional database for estimating facilities and wells decommissioning costs. Consultancy companies such as Genesis Oil and Gas Consultants, with a global centre of excellence based in the United Kingdom, have specialized in determining the appropriate decommissioning option and providing detailed technical and cost studies including environmental impact assessment, risk assessment and safety case analysis. For well decommissioning costs, the IHS Markit Rushmore Drilling Performance Review provides a useful benchmark for cost estimation by reporting global operators' actual well P&A (plug and abandonment) costs as an analogue (Ref. [12]).

To provide some perspective on decommissioning costs, a UKCS subsea single well decommissioning cost is in the range $10–20 m, depending on complexity of the well and water depth. In 2017 Shell decommissioned the Brent Delta platform, a concrete gravity-based structure, and the total cost reported was $650 m, with 40% of this spent on the P&A of the wells. The Allseas *Pioneering Spirit* was used to remove the 24,500 tonnes of topsides, which were returned to shore, while the concrete base remains in situ. A report of the activities was provided by Shell (Ref. [13]) with final costs reported to OFRED but not included in the report. The cost estimate above was reported in an industry publication (Ref. [14]).

When decommissioning cost was introduced as a cash flow item in Section 7.4 it was entered as a single cost in one single year. While this may be adequate for incorporating into the economic model for making development planning decisions, where the cost is some 10–20 years (the economic lifetime) away from the reference date, the actual decommissioning activities and costs will not be contained to one year. If evaluating the remaining NPV in a late-life asset, the timing of the decommissioning costs will become more important, and the phasing of costs should be incorporated according to the activity plan. Some equipment may be decommissioned prior to CoP, in preparation, and may form part of the annual opex. The main costs will be incurred shortly after CoP, but there will be residual costs for post-CoP monitoring, maintenance and management.

9.4.4 Fiscal treatment of decommissioning costs

The responsibility for decommissioning activities rests with the operator at the point of CoP. The cost of the activities rests with the owners of the asset at this point. However, depending on the fiscal terms, the decommissioning costs may be claimed as a tax deduction against revenue. Detailed understanding of the taxation system specific to the host country is required to calculate the net-of-tax cost of decommissioning.

Under the legislation for oil and gas taxation in place in 2020 in the United Kingdom, as a specific example, decommissioning cost can be claimed against an asset's previously paid tax under a loss carry back

(LCB) arrangement. The UK Government is therefore compensating the owners for decommissioning, to some extent. Table 9.7 illustrates the mechanism for LCB for a single ring-fenced asset which has paid Supplementary Charge (SC) and Corporation Tax (CT) during its producing lifetime, but has been decommissioned in Year 20X8. The operator cannot of course claim the tax repayment until the decommissioning costs are spent, so in principle, the repayment is the product of the decommissioning cost multiplied by the tax rate. In the example, the SC rate changes, as it has done in the United Kingdom (refer to Fig. 6.11). The calculation therefore must work backwards from the date of the decommissioning expenditure. Both CT and SC are ring fenced to the specific field, so the tax refund relates only to taxes previously paid on that field.

Table 9.7 Example of loss carry back of decommissioning costs.

YEAR		20X1	20X2	20X3	20X4	20X5	20X6	20X7	20X8
GROSS REVENUE	$m	500	400	300	200	100	50	20	
CAPEX	$m		50						
OPEX	$m	100	100	100	100	50	25	10	
ORIGINAL TAXABLE INCOME	$m	400	250	200	100	50	25	10	
CT + SC TAX RATE		50%	50%	50%	40%	40%	40%	40%	
ORIGINAL CT + SC TAX	$m	200	125	100	40	20	10	4	
DECOMMISSIONING COST	$m								400
IN-YEAR RELIEF	$m								N/A
RESTATED TAXABLE INCOME	$m	400	235	0	0	0	0	0	
CUMULATIVE LOSS CARRY BACK	$m		400	385	185	85	35	10	
REFUND OF TAX PAID	$m		7.5	100	40	20	10	4	181.5

The decommissioning cost is incurred in Year 20X8, but there is no in-year tax relief since there is no taxable income. Working back from that date the restated taxable income is set to zero and the cumulative LCB is calculated backwards until the cumulative LCB equals the total decommissioning costs. Care is required in the Year 20X2 (in this example) to adjust the restated taxable income so that the cumulative LCB equals the decommissioning cost.

The refund of tax paid is then the original taxable income less the restated taxable income, multiplied by the relevant tax rate for that year. In this example, the total refund is $181.5 m for the decommissioning cost of $400 m. The only reason that this is not a very simple calculation is that the historic total tax rate has changed during the LCB calculation. In fact, SC has varied in the United Kingdom from 32% to 20% to 10%. Some fields are also liable to Petroleum Revenue Tax (PRT), and in these cases, the tax refund can also include PRT payments. In the case of multiple fields, cross-field allowances and other investment allowances, the tax calculation becomes more complex.

Details of this mechanism are neatly summarised by Deloitte in Ref. [15] and by Oil and Gas UK for the UK-specific fiscal arrangements in Ref. [16].

As a comment on the accounting treatment of decommissioning costs, the accountant will usually place a series of costs in the annual profit and loss accounts as a provision for the final decommissioning. This follows an accounting principle of 'prudence' and avoids a large accounting loss in the year of actual decommissioning. This provision is not allowed as an expense against the CT described in the example, which is in fact called Ring Fenced Corporation Tax (RFCT) in the United Kingdom, as this form of CT is specific to the ring-fenced asset.

9.4.5 Decommissioning cost liabilities

If an asset in the UKCS is sold, the new owner is able to claim the tax refund based on the tax paid by the previous owner. This means that, regardless of the ownership at the point of decommissioning, the UK government is providing the tax refund by applying the historic tax rates. This will be factored into an evaluation of a late-life asset purchase and reduces the liability of the new operator. The UK fiscal system allows this transfer of tax refunds to avoid discouraging new owners from purchasing late-life assets.

Nevertheless, the new owner has to be in a position to fund the balance of the decommissioning cost, $218.5 m in the example. In any region, this may pose a concern to the host government if there is a risk that the owners are unable to fund the activity.

This issue was brought into focus in 2005 in the United Kingdom when Tuscan Energy, the operator of Ardmore field, a redevelopment of the original Argyll field, declared insolvency and was unable to fund the decommissioning of the facilities, nearly leaving the UK tax payer with the liability.

To ensure that the tax payer is not exposed to an unacceptable risk, the host government can take several approaches. In a concessionary system (refer to Fig. 6.8), the host government will hold the owners of an installation jointly and severally liable for the decommissioning and any subsequent inspection and maintenance costs. If one of the joint venture owners is unable to meet the liability, it is passed to the remaining partners, and this obligation remains in perpetuity.

As a further power, in case of default by the asset owners at the time of decommissioning, the government can roll back the liability to former owners, thus protecting the tax payer. In the sale of an asset, the seller therefore needs to have some confidence that the buyer will be able to cover the decommissioning costs, or else the seller is also exposed. The potential for default payments has given rise to the introduction of decommissioning security agreements (DSAs) as a commercial risk management tool.

DSAs are a commercial arrangement, whereby a third party runs a trust set up to handle funds between two parties involved in a decommissioning obligation. For example, this can be between a seller and a buyer of an asset, and the fund will provide the security to cover any shortfall in the ability of the buyer to meet the obligation. The security can take several forms, such as a parent company guarantee (PCG), a Letter of Credit (LoC) or cash. PCGs are suitable only for larger companies (as the buyer), and for smaller companies the LoC is often the preferred solution. The LoC will be issued by a highly rated bank on behalf of the buyer and will be for a sum equivalent to any anticipated shortfall in meeting the obligation. Both the seller and the government need to be satisfied with the trustworthiness of the bank issuing the LoC. In effect, the LoC acts as insurance for the seller and the government, and of course, the buyer pays a premium for this.

In the United Kingdom, some of the larger companies, who have now exited the UKCS and have sold assets to smaller companies, have used LoCs to cover their risk of having the decommissioning cost rolled back to them should the small new owner default. The sum guaranteed in the LoC will vary each year depending on the current estimate of decommissioning costs (which reduces as technology and methodology advance) and the remaining value of the small company which is linked to the remaining NPV of the asset (and reduces as remaining reserves decline). Both the costs and the value are of course a matter of opinion, and they are often resolved by referral to an expert. For a legal perspective on DSAs, Ref. [17] is recommended. It describes the original joint venture partners in the asset as first-tier participants, being those who are at risk of being caught by the rollback of liability created by second-tier participants who have bought the late-life asset.

In a contractual arrangement such as a PSC (refer to Fig. 6.8), the oil company acts as a contractor operating on behalf of the host government. The contract includes a production period during which the operator oil company is entitled to cost oil recovery and a share of profit oil. If the CoP date lies beyond the production period, unless the contract is extended, the asset will become 100% owned by the government, usually represented by the national oil company (NOC) which then continues the production operations. This time is likely to be approaching the EL of the field, which will then require decommissioning.

To protect the NOC from the full decommissioning cost, the contract will typically require a portion of the gross revenues earned during the production period to be directed into a decommissioning fund (formerly an abandonment fund), held in escrow for the specific purpose of paying for the final activities. A third party manages an escrow account on behalf of the other two parties, only releasing the funds on the appropriate instruction. This arrangement is a legal concept and provides protection in the same way as a LoC.

Cameron and Stanley (Ref. [18]) report in 2017 that recent PSCs in Malaysia use a specific methodology for determining the amount contributed to the decommissioning fund. The operator is asked to make an estimate of the final decommissioning cost, and then a factor related to the remaining reserves is applied to determine how much cost oil recovery can be claimed in any one year. That sum is placed into a decommissioning account.

Even if the CoP falls within the contractual production period due to reaching the EL, there will be no cost oil against which the oil company can claim an allowance, and within the PSC area there will by definition be no net cash flow. An abandonment fund is thus required to ensure decommissioning can be carried out.

Grey areas can arise around the legal liabilities for decommissioning the infrastructure constructed prior to the handover to the NOC, and that installed thereafter. For a legal view on how oil companies and governments can protect their interest in settlements concerning decommissioning liability, Ref. [19] provides guidance.

References

[1] J.P. Martins, J.M. MacDonald, C.J. Stewart, C.J. Phillips, The Management and Optimization of a Major Wellwork Program at Prudhoe Bay, SPE-30649-MS, 1995.
[2] M. Sidahmed, E. Ziegel, S. Shirzadi, D. Stevens, M. Marcano, Enhancing Wellwork Efficiency with Data Mining and Predictive Analytics, SPE-167869-MS, 2014.

[3] UK Oil and Gas Authority (OGA), Guidance on Requirements for the Planning for Cessation of Production, August 2019. https://www.ogauthority.co.uk/media/6033/oga_cop_external_guidance_aug_19.pdf.
[4] Rystad Energy, Cost Analysis, April 2020. http://www.rystadenergy.com.
[5] SPE, Petroleum Resources Management System, 2018, ISBN 978-1-61399-660-7. https://www.spe.org.
[6] UK Department for Business, Energy & Industrial Strategy, Guidance Notes for Decommissioning of Offshore Oil and Gas Installations and Pipelines, November 2018. https://assets.publishing.service.gov.uk/government/uploads/system/uploads/attachment_data/file/760560/Decom_Guidance_Notes_November_2018.pdf.
[7] Oil & Gas Authority, UK National Data Repository. https://ndr.ogauthority.co.uk/.
[8] Oil & Gas UK, Decommissioning Insight, 2019. https://oilandgasuk.co.uk/wp-content/uploads/2019/11/OGUK-Decommissioning-Insight-Report-2019.pdf.
[9] Oil & Gas Authority, UKCS Decommissioning, 2019 Cost Estimate Report, July 2019. www.ogauthority.co.uk.
[10] IHS Markit, reportDecommissioning Quarterly Report. https://ihsmarkit.com/products/decommissioning-quarterly-report.html.
[11] https://allseas.com/equipment/pioneering-spirit/.
[12] IHS Markit, Rushmore Drilling Performance Review (DPR). https://ihsmarkit.com/products/drilling-performance-review.html.
[13] Shell U.K, Limited, Brent Delta Topsides Decommissioning Close-Out Report, December 2019. https://www.shell.co.uk/sustainability/decommissioning/brent-field-decommissioning/brent-field-decommissioning-programme/.
[14] Rigzone. https://www.rigzone.com/news/north_sea_decommissioning_costs_on_the_rise-19-jun-2019, June 2019.
[15] Deloitte, Oil and Gas Taxation in the UK, Deloitte taxation and investment guides, 2020. https://www2.deloitte.com/content/dam/Deloitte/global/Documents/Energy-and-Resources/dttl-er-UK-oilandgas-guide.pdf.
[16] Oil & Gas UK, Decommissioning, A Tax Guide, A Summary of the Fiscal Relief Available within the UK Tax Regime, October 2015.
[17] Squire Patton Boggs, Decommissioning in the UK Continental Shelf: Decommissioning Security Disputes, 2016 International Energy Law Review, I.E.L.R., 2016.
[18] P.D. Cameron, M.C. Stanley, Oil, Gas, and Mining: A Sourcebook for Understanding the Extractive Industries, World Bank, Washington, DC, 2017, https://doi.org/10.1596/978-0-8213-9568-2. License: Creative Commons Attribution CC BY 3.0 IGO.
[19] Quinn Emanuel Urquhart & Sullivan LLP, Exiting Long-Term Contracts with Governments: Trends in Production Sharing Agreement Disputes in the Oil & Gas Sector, December 2019. https://www.jdsupra.com/legalnews/december-2019-exiting-long-term-50170/.

CHAPTER 10

Portfolio management

Contents

10.1 Managing value and risk	309
10.2 Portfolio effect on volumes and risk	310
10.3 Portfolio effect on value and risk	312
10.3.1 Combining assets parametrically	315
10.3.2 Combining assets using Monte Carlo simulation	318
10.3.3 Systematic and unsystematic portfolio risk	320
10.3.4 Markowitz diversification	321
10.4 The efficient frontier – balancing the portfolio to match strategy	324
10.5 Gambler's ruin and exploration portfolios	330
References	335

10.1 Managing value and risk

The chapters so far have focused on the economics and risk of single assets, be they exploration prospects or development projects. Value indicators (NPV, EMV), efficiency indicators (IRR, NPV/PVCapex) and risk indicators (s.d./mean of NPV) have all been calculated on a stand-alone asset basis. When assets are combined in a portfolio, some properties of the individual asset are additive, others are not. This chapter will provide an overview of how investing in a portfolio of assets affects value and risk.

The value of a portfolio of combined assets can be calculated by simply adding the NPV of each asset, weighted by the working interest in that asset. NPV is an additive parameter. Risk, however, is not additive, and calculating the risk of a portfolio requires care, but is a key attraction of building a portfolio. Portfolio management should not be simply a matter of adding the value of a collection of assets; it should also calculate the associated impact on risk.

The saying 'don't put all your eggs in one basket' implies that diversifying one's investment capital across a number of assets reduces risk. As a matter of historical interest, the phrase was used by Nobel Prize winner for Economics, Professor James Tobin of Yale University when asked to summarise many years' of his labour focused on portfolio choice, stating that 'his work showed that it was not wise to put all your eggs in one basket'. Though the origin of the phrase is debated, it is commonly attributed to Miguel Cervantes, who wrote Don Quixote in 1605. The concept clearly has deep roots.

This chapter will discuss why the phrase is true, and show how reduction in risk can be achieved by combining assets that are either correlated or uncorrelated to each other. The efficient frontier plot introduced in Section 8.6.5 will be revisited to demonstrate that a portfolio choice should aim to deliver the maximum attainable value for a target level of acceptable risk, or to deliver the target value at the minimum attainable level of risk. The financial adviser's definition of risk, being the standard deviation of value divided by the mean value (σ/μ), is again used throughout this chapter, though it is noted that some texts define the risk simply as the standard deviation of the value (σ). The (σ/μ) is actually the coefficient of variation (CV) and can be considered as a method of normalizing the measure of risk so that it becomes a dimensionless ratio rather than a value.

10.2 Portfolio effect on volumes and risk

As an introduction, consider that a company has the opportunity to invest in developing up to seven independent oil field discoveries. Each has a volumetric range of recoverable oil, which is log-normal in shape, and the volume is represented by a mean and standard deviation (s.d.), as shown in Table 10.1.

Table 10.1 Ranges of uncertainty in recoverable oil for seven discoveries.

Discovery	Short name	Mean volume (MMb)	s.d. volume (MMb)	s.d./mean (= risk)
Radon	R	100	30	0.30
Strontium	S	300	110	0.37
Titanium	T	50	10	0.20
Uranium	U	200	60	0.30
Vanadium	V	500	175	0.35
Wolfram	W	250	100	0.40
Xenon	X	400	180	0.45

Suppose the corporate objective is to add 500 MMb of recoverable oil. This could be achieved by developing the Vanadium discovery. Using the financier's definition of risk as s.d./mean, (σ/μ), which as noted is the coefficient of variation (CV), Vanadium has a relatively high risk.

An alternative would be to create a portfolio of Strontium (S) and Uranium (U), which would also deliver 500 MMb. The σ/μ of this portfolio could be assessed through Monte Carlo simulation by defining the distribution of the two inputs, defining the sum as an output, running the simulation and then extracting the portfolio mean (μ) and s.d. (σ), and hence risk (σ/μ).

However, since the inputs are distributions that are represented by μ and σ values, Monte Carlo simulation can be replaced by a parametric method, using the following statistical rules that apply to normal and log-normal distributions. The independent assets are A and B, and the portfolio is represented by A + B.

$$\mu_{(A+B)} = \mu_A + \mu_B \tag{10.1}$$

$$\sigma^2_{(A+B)} = \sigma^2_A + \sigma^2_B \tag{10.2}$$

Recall that σ^2 is the variance. The risk (σ/μ) of the portfolio of Strontium and Uranium (S + U) is calculated to be 0.25, and is plotted along with the individual discoveries in Fig. 10.1. This shows that the portfolio (S + U) delivers the same volume as Vanadium (V) but for a lower level of risk, despite the fact that S has a higher level of individual risk than V. Since an objective of portfolio management is to deliver the corporate target (volume in this case) at the minimum risk, the portfolio (S + U) is a better option than putting all one's eggs into the Vanadium basket.

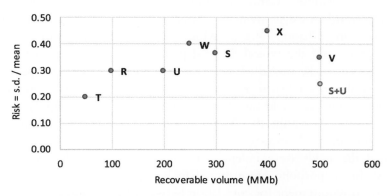

Figure 10.1 Risk vs volume plot for seven independent assets R–V.

You may have noted that another option would be to develop Wolfram (W), Uranium (U) and Tritium (T), which would also deliver 500 MMb. The risk of this portfolio is 0.23, which is better still. Radon plus Xenon also achieves the target, with a risk level of 0.36, much the same as Vanadium alone.

More combinations can be created by changing the working interest in the assets. To do this the asset parameters (μ, σ) in Eqs. (10.1) and (10.2) must be weighted by the fractional working interest in each asset. In the E&P industry, opportunities to farm-in and farm-out of assets both domestically and internationally are abundant, so the number of combinations of individual assets in a portfolio that can still achieve a target is almost limitless.

Other factors will constrain the number of feasible combinations, such as the total capital required to fund each combination, or the work force required to develop and operate multiple fields. The latter part of Section 10.4 addresses the application of such constraints.

This simple example has used volume as the target for the portfolio, but the target will more commonly be set in economic terms, such as achieving a portfolio internal rate of return (IRR) or value (NPV). The example also assumed that the assets were independent, meaning that the volume of one asset did not have any correlation with another. The following sections consider economic targets and covariance of assets.

10.3 Portfolio effect on value and risk

Generating a continuous probability distribution of hydrocarbon volumes using the Monte Carlo technique is common and relatively straightforward, since it involves a limited number of input variables as described in Chapter 4.2.2. Most of these input variables are independent, and the typical linear dependency between, say, saturation and porosity can be modelled using the convenient correlation options in commercial software such as @RISK or Crystal Ball. The continuous distribution for each field can then be combined in a portfolio by adding the distributions within the Monte Carlo simulation or by using the parametric method shown in the example in Section 10.2, which assumed the fields were independent. Generally, the volumes in each field within a portfolio are assumed to be independent, in which case there is no requirement for correlation between field volumes.

While the probabilistic approach to volumetric estimation is long established within the E&P industry, it was slower to adopt a similar approach to representing uncertainty in value, often preferring to select several specific volumetric outcomes (e.g. P90, P50 and P10) and evaluating each volumetric case along with the development concept to yield discrete NPVs. Swanson's rule (see Section 4.2.2) can then be applied to weight the three outcomes and calculate the EMV of the project, being the probability-weighted outcome, or average. This is the approach taken in all the decision tree examples in this book so far, but it only shows three discrete outcomes of value and the average, and neither does it provide the probability of achieving the average.

To move towards a probabilistic distribution of NPV, it is possible to apply Monte Carlo simulation in the economic model. Distributions can be attached to the input variables, such as oil price, capex, opex, as shown in Fig. 10.2, using a triangular distribution for oil price as an example. The spreadsheet cell in the economic model containing the project NPV is defined as a Monte Carlo output. Alternatively, a distribution for inputs can be achieved by attaching a distribution to the multiplier values MULT used in the sensitivity analysis in Table 8.7.

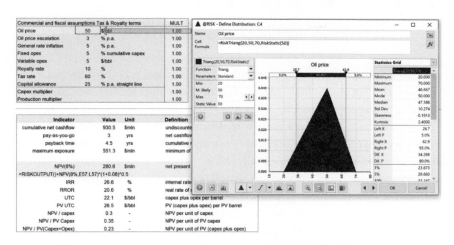

Figure 10.2 Economic model with input variables as distributions.

Any input can have a distribution attached to it, and inputs can be correlated using the correlation option in @RISK. Positive and negative correlation coefficients can be handled. When run, the output will be a fully probabilistic distribution on NPV, shown in Fig. 10.3 as a reverse cumulative probability of the NPV(8) in $m. The left delimiter on the graph has been adjusted to show that there is an 8.5% cumulative probability of a zero NPV(8) in this project, i.e. an 8.5% chance of being an economic failure The P90 NPV is just positive ($21.2 m), as can be read from the statistics grid, noting that @RISK is using a reverse cumulative probability notation.

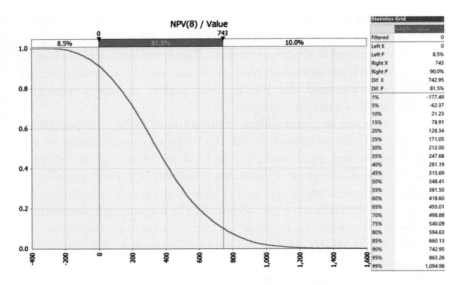

Figure 10.3 Probabilistic distribution of NPV.

However, there is a significant web of non-linear relationships between many of the inputs, which can undermine the confidence placed in the simple spreadsheet approach.

For example, as the volume of hydrocarbon initially in place (HCIIP) varies, so may the development concept. This will not be a linear relationship as there will be break points on the volume scale at which one concept such as a subsea tie-back switches to a different concept such as a fixed production platform. As different volumes are randomly generated in the Monte Carlo simulation, each specific volume should be linked not only to a development concept and its cost but also to the recovery factor, production profile, capex, opex and possibly the fiscal system.

Particularly in contractual fiscal systems such as PSCs, the government take can vary with production level and with oil or gas price. Certain production levels or commodity prices may trigger changes in royalty, or the split of the profit oil, or even one-off production bonuses payable to the government. If these relationships are part of the agreed fiscal terms, they should also be coded into the economic model.

The web of dependencies is extensive. This level of complexity partly explains why the economic evaluation has traditionally been based on selected volumes such as the P90, P50 and P10.

Although significant effort is required to generate a faithful probabilistic NPV curve such as Fig. 10.3, the result is very appealing. It shows the cumulative probability of failure, and the probability of achieving the mean NPV, which decision tree analysis does not. The EMV on a decision tree is the probability-weighted average of a number of discrete outcomes, but the probability of achieving this is not apparent.

If each asset can be represented by a distribution curve of NPV by using either the Monte Carlo technique or a parametric approach, the next step is to combine the assets together into a portfolio to assess combined NPV and risk. While correlations between input variables for the NPV of each asset has been discussed, it may also be that assets have a covariance with each other, adding a further level of complexity, which will be considered in Sections 10.3.1 and 10.3.2.

10.3.1 Combining assets parametrically

Assuming that each asset value can be represented by a normal or lognormal distribution using a mean (μ) and standard deviation (σ), and that the asset values are not correlated, it is relatively simple to combine them to establish the portfolio value and risk, in a similar fashion to dealing with the volumes in Section 10.2.

If it is felt that the assets or projects are correlated, then the parametric method becomes more complicated, but still possible. The mean NPV of the portfolio is still the sum of the mean of each asset. Assuming there are n assets in total and i represents a single asset with a mean NPV of μ_i, the mean NPV of the portfolio μ_p is given by

$$\mu_p = \sum_{i=1}^{n} \mu_i \qquad (10.3)$$

The standard deviation of the portfolio σ_p can be calculated by

$$\sigma_p = \sqrt{\sum_{i=1}^{n} w_i^2 \sigma^2(k_i) + \sum_{i=1}^{n} \sum_{j=1}^{n} w_i w_j \text{Cov}(k_i k_j)} \qquad (10.4)$$

where
  n = number of assets in the portfolio
  σ_p = standard deviation of the portfolio of assets
  w_i = proportion of the ith asset in the portfolio
  w_j = proportion of the jth asset in the portfolio
  $\sigma^2(k_i)$ = variance of the ith asset
  $\text{Cov}(k_i\ k_j)$ = covariance of the ith asset and the jth asset

If there is no covariance between the assets and 100% of each asset is in the portfolio, Eq. (10.4) reduces to the simple form in Eq. (10.2).

The covariance between the asset values is calculated from the correlation coefficient between them. This requires data to analyse, with pairs of responses for each asset. Most statistics texts such as Refs. [1,2] describe the procedure for calculating the correlation coefficient between two variables, and this will be summarised here.

Fig. 10.4 is a demonstration of a strong and weak correlation between pairs of values, a conclusion that simple inspection could judge qualitatively.

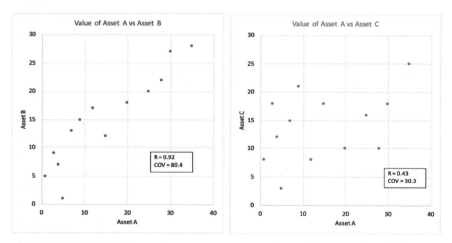

Figure 10.4 Correlating variables using Excel functions = CORREL and = COVARIANCE.

The correlation coefficient (R) is a normalised version of the covariance and is defined as

$$R = \frac{Cov(k_i k_j)}{\sigma(k_i)\sigma(k_j)}$$

A correlation coefficient (R) of 1.0 would be a perfect positive correlation, 0.0 no correlation and -1.0 a perfect negative correlation.

Calculating R allows the covariance between the assets to be calculated. The covariance between each pair of assets in the portfolio must be calculated to satisfy Eq. (10.4), which can be time-consuming if done by hand. Excel provides a useful function = CORREL(array1,array2) to calculate R for any pair of assets.

Excel also has a function = COVARIANCE(array1,array2) that calculates the covariance between pairs of assets. This speeds up the process considerably, but even then, the covariance between each pair of assets needs to be calculated and added together in Eq. (10.4).

Using the Excel formulae for convenience, the data in Fig. 10.4 are computed to have a strong correlation ($R = 0.92$) between Asset A and Asset B, but a weak correlation between Asset A and Asset C ($R = 0.43$). Unless correlation is significant, it is not worthwhile including in the model. Generally, a value of R greater than 0.7 is considered to represent strong correlation. A value of R between 0.5 and 0.7 is considered a moderate correlation, and anything less than 0.4 is considered weak or no correlation. On this basis, unless R is at least 0.7 the correlation is probably not worth including in the portfolio model.

With data such as that shown in Fig. 10.4, and with Excel on hand, it is straightforward to calculate R and the covariance between any two assets, and thus build up the portfolio standard deviation σ_p using Eq. (10.4). The portfolio risk, as defined in this book, is then σ_p/μ_p.

As Eq. (10.4) shows, the more pairs of assets that have a positive covariance, and the higher the positive value of the covariances, the higher the standard deviation of the portfolio value, and hence the higher the portfolio risk compared to there being no covariance between assets. If pairs of assets have a negative covariance (i.e. a negative correlation coefficient), this will reduce the portfolio risk. This point will be revisited in Section 10.4 when considering the balancing of a portfolio and the influence of including negatively correlated assets in a diversified portfolio.

The challenge in describing correlation between asset values is to source the data that provides the input. Financial investors have the benefit of a history of share prices and can plot the instantaneous price of share A against the price of share B at many points in time, providing a convenient data set from which to calculate the covariance. Data sets to estimate asset values are harder to generate in the oil and gas industry, but the single most common global denominator that influences value is the oil or gas price.

The value of oil field assets will be strongly correlated through the dependence on oil price, unless hedging has been practised to sell oil at an agreed price in the future. Gas prices are also linked to oil price, but the correlation will be weaker, especially if long-term gas sales contracts have been agreed at prices that are only partially indexed to oil price.

If a portfolio is composed of assets within the same country, then changes in the local fiscal system will create a significant correlation between asset values. If assets are spread across geographic regions, any correlation due to this driver will be weak.

One approach to estimating the correlation or covariance between asset values is to take a scenario approach to forecasting common variables such as oil price. One scenario may envisage long-term oil price decline as demand reduces due to increased energy supply from renewable resources. Since oil is a commodity item, this will affect all oil assets and the correlation between the values of oil assets will be strong, while the correlation between the value of gas assets and oil assets will be positive, but weaker.

Multiple scenarios can be created to generate discrete portfolio values under each scenario. Royal Dutch Shell implemented scenario planning in the 1970s, and Refs. [3,4] provide a history and progression of the approach, stressing that scenarios should be plausible, not extreme, and that they need to be kept updated. As the energy industry transitions from traditional fossil fuels to renewables, the latter point is very pertinent.

10.3.2 Combining assets using Monte Carlo simulation

If each asset value can be described with a probability distribution in Excel, then Monte Carlo simulation tools such as @RISK or Crystal Ball can add the asset value distributions together. The formula in the cell describing the portfolio is simply the sum of the assets. When the simulation is run, the asset values can be independent, i.e. non-correlated, or correlated together using the correlation matrices available in the software. Fig. 10.5 shows the input for four assets, three of which are correlated, with formulas displayed to illustrate how the sum of the assets is entered as a portfolio. The software also checks the internal consistency of the correlation coefficients, and adjusts accordingly.

ASSET	@RISK input of NPV ($m)
Asset A	=RiskLognorm(150,30,RiskStatic(150),RiskCorrmat(Portfoliocorrelations,1))
Asset B	=RiskLognorm(200,50,RiskStatic(200),RiskCorrmat(Portfoliocorrelations,2))
Asset C	=RiskLognorm(80,30,RiskStatic(80),RiskCorrmat(Portfoliocorrelations,3))
Asset D	=RiskLognorm(450,100,RiskStatic(450),RiskCorrmat(Portfoliocorrelations,4))
Portfolio	=RiskOutput()+SUM(C4:C7)

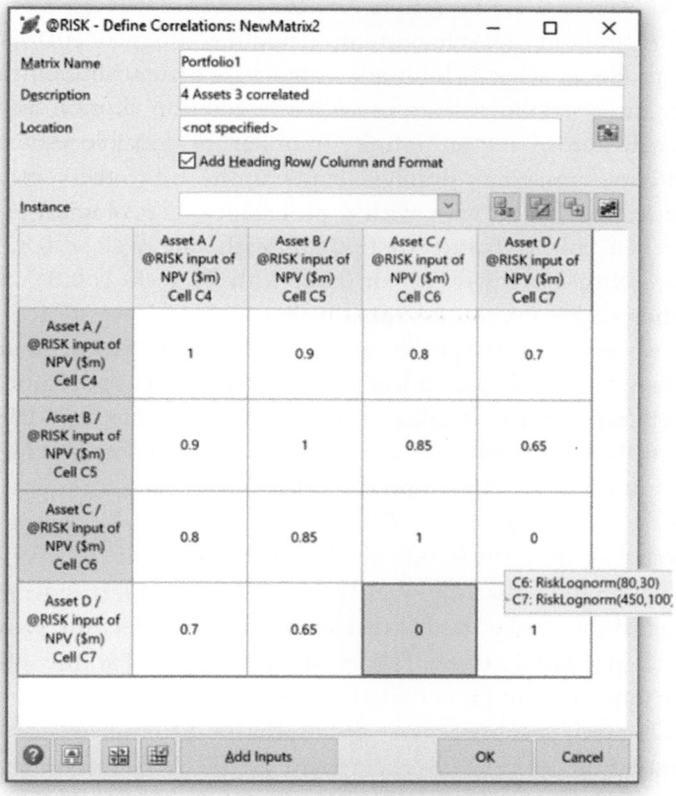

@RISK - Define Correlations: NewMatrix2

Matrix Name: Portfolio1
Description: 4 Assets 3 correlated
Location: <not specified>

	Asset A / @RISK input of NPV ($m) Cell C4	Asset B / @RISK input of NPV ($m) Cell C5	Asset C / @RISK input of NPV ($m) Cell C6	Asset D / @RISK input of NPV ($m) Cell C7
Asset A / @RISK input of NPV ($m) Cell C4	1	0.9	0.8	0.7
Asset B / @RISK input of NPV ($m) Cell C5	0.9	1	0.85	0.65
Asset C / @RISK input of NPV ($m) Cell C6	0.8	0.85	1	0
Asset D / @RISK input of NPV ($m) Cell C7	0.7	0.65	0	1

C6: RiskLognorm(80,30)
C7: RiskLognorm(450,100)

Figure 10.5 Excel portfolio summation and correlation of asset values in @RISK.

Moving beyond the scope of Excel modelling, some E&P companies have pursued the development of internal economic models to produce the probabilistic NPV, as have several commercial enterprises, including FOCUS from Indeva (Ref. [5]), Merak PEEP from Schlumberger (Ref. [6]), PetroVR and Aucerna Portfolio from Aucerna (Ref. [7]).

These more complex models contain the detail of the fiscal systems in different countries, providing the ability to create portfolio value and risk for a geographic spread of assets.

10.3.3 Systematic and unsystematic portfolio risk

Portfolio investments reduce risk through diversification, whether the individual assets are correlated or not.

Investing in multiple non-correlated (i.e. independent) assets of similar size produces what is known as simple or natural diversification. In this case, the portfolio risk drops with the addition of each asset, and tends towards zero with an infinite number of independent assets in the portfolio. Imagine drilling exploration oil wells in geologically independent areas, each with a probability of geological success (P_g) of 0.2. The risk of failure to find oil with one well is 0.8. The risk of failure with 5 wells is 0.8^5 or 0.32, with 20 wells is 0.8^{20} or 0.011, and tending to zero with additional wells.

However, in the E&P sector, asset values are typically correlated through oil and gas prices, being the most significant commercial uncertainties in the business, as demonstrated in Fig. 6.3. When assets are positively correlated (e.g. a change in oil price moves the value of all assets in the same direction), the risk tends towards a residual risk level, as illustrated in Fig. 10.6. This level of risk is typically reached after diversifying into 5–10 individual assets.

The residual risk is known as undiversifiable risk or systematic risk, and is due to the market factors that simultaneously affect the value of all assets in the portfolio. These factors are usually outside the control of the company, and can be considered as external risks.

Oil and gas prices are prime causes of undiversifiable risk from investing in a portfolio exclusively made up of upstream oil and gas assets. Other techniques such as hedging forward price must be applied to address price-related risk. Other general sources of undiversifiable risk include inflation, exchange rates - which hedging can also reduce, and general recession or war, against which it is very difficult to protect any business.

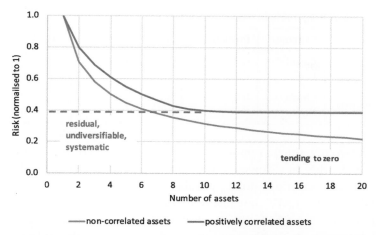

Figure 10.6 Example of risk vs number of correlated and non-correlated assets.

In the E&P business, portfolio management is primarily trying to address unsystematic risk, since this is influenced by the factors that are under the control of the company (e.g. project cost, schedule, reservoir recovery factor), which can be considered as internal risks.

Diversifying the portfolio into assets that have weak correlations is a method of addressing the unsystematic risk. This method was introduced by Harry Markowitz in 1959, and is known as Markowitz diversification (Ref. [8]).

10.3.4 Markowitz diversification

Markowitz diversification depends upon creating a portfolio composed of assets that have a correlation coefficient less than 1.0. The lower the correlation, the more effective the method.

By way of example, consider the opportunity to invest in two industries, car manufacturing and pharmaceuticals. In the first instance, you choose to place all your investment capital in just one of these. The forecast NPV of your investment shown in Table 10.2 depends on the future performance of each industry, with a 50/50 probability of each being good or poor. The Expected Monetary Value (EMV) of the investment made in either pharmaceuticals or motors has the same value ($22), so you should be ambivalent about which industry to invest in. In both investments, the actual outcome will be either $32 or $12, but never $22, remembering that the EMV is the probability-weighted outcome, or the mean outcome (μ). The standard deviation of the population (σ) can be calculated in Excel to be 10, and hence the risk (σ/μ) is 0.45.

Table 10.2 Investing 100% in either one of two non-correlated industries.

Investment	Outcome	Probability	NPV($)	EMV($)	s.d. ($)	Risk
100% cars	Good	0.5	32	22	10	0.45
	Poor	0.5	12			
100% pharma	Good	0.5	32	22	10	0.45
	Poor	0.5	12			

With two investment opportunities, it is possible to create a portfolio. In Table 10.3, the investment capital is split 50/50 between the two opportunities, creating four possible outcomes, each with 25% probability. The EMV remains the same, but now there is a 50% probability of actually realising the EMV of $22 and only a 25% probability of realising the extremes, which reduces the risk to 0.32, again calculated in Excel (by inputting 25 values of 32, 25 values of 12 and 50 values of 22 and using = AVERAGE and = STDEV functions).

Table 10.3 Investing 50% in each of two non-correlated industries.

Investment	Outcome	Probability	NPV($)	EMV($)	s.d. ($)	Risk
50% cars	Good, good	0.25	32	22	7.1	0.32
	Good, poor	0.25	22			
50% pharma	Poor, good	0.25	22			
	Poor, poor	0.25	12			

In the cars and pharmaceutical example, it was assumed that there was no correlation between the assets in the portfolio, and risk was reduced by the portfolio investment compared to putting all eggs in one basket. The weaker the correlation between assets, the more the portfolio risk is reduced, and so if one could find assets that are negatively correlated then this has a more significant impact in reducing portfolio risk.

In the following example, again taken to the extreme, the opportunity is to buy shares in two companies, one manufactures raincoats and the other sunglasses. The performance over the next summer season depends on the weather, an uncontrollable factor, which is forecast to be bright and sunny or dull and wet, with a 50% probability of either outcome. If either industry does well, the outcome of shares for each is an NPV of $40, but if either does poorly, the NPV is $10.

Table 10.4 shows that investing 100% in one asset yields a finite risk, in this case 0.6, but by diversifying into two perfectly negatively correlated assets the risk is reduced to zero. The $25 EMV is in fact the NPV, as there is a 100% probability of achieving this; risk has been eliminated.

Table 10.4 Diversification with perfectly negative correlation between assets.

Investment	Weather	Probability	NPV($)	EMV($)	s.d. ($)	Risk
100% sunglasses	Bright hot	0.5	40	25	15	0.6
	Rainy	0.5	10			
100% raincoats	Bright hot	0.5	10	25	15	0.6
	Rainy	0.5	40			
50% sunglasses	Bright hot	0.5	50	25	0	0
50% raincoats	Rainy	0.5	50			

This example is of course extreme, but makes a point. In practice, it is unlikely that assets will have a perfectly negative correlation ($R = -1.0$), but if it is possible to include assets into the portfolio that have a significant negative correlation then they will very effectively reduce risk. Financial investors seek such opportunities. Arguably, in the first example of cars and pharmaceuticals, there may be a negative correlation between the values of these industries due to external factors. During the Covid-19 pandemic, the demand for new cars dropped dramatically while demand for pharmaceutical products rose.

In the E&P business, oil and gas price represent the largest uncertainty in the business, leaving a residual or undiversifiable risk if investment is restricted to the upstream sector. However, by diversifying the business and investing in both upstream and downstream (refining and chemicals) sectors, the risk associated with oil price can be significantly reduced. Low oil price is bad for the E&P business but good for the downstream business where the feedstock cost becomes lower. This diversification is one of the advantages of a vertically integrated company that makes a strategy decision to spread its interests across both market sectors.

As the energy industry enters a transition from traditional fossil fuels to renewable sources, diversification from oil and gas E&P into renewables is a clear opportunity for diversification, which many major oil companies have built into their investment strategies. For example, the rebranding of

Norway's Statoil to Equinor in 2018 heralded its shift from a pure oil and gas company to an energy provider by introducing offshore wind and solar to its portfolio.

Other risks in the E&P business can be reduced through diversification, as suggested in Table 10.5.

Table 10.5 Risk reduction through diversification options.

Source of risk	Diversification options
Oil price	Investment in upstream and downstream business
Gas price	Multiple assets with different long-term gas sales contracts
Inflation and exchange rate	Investment in different countries
Fiscal terms	Investment in different countries
Political risk	Investment in different countries
Reserves	Investment in multiple oil and gas fields
Exploration success	Blend of high risk high reward and low risk low reward prospects
Cost	Investment in multiple oil and gas fields

10.4 The efficient frontier — balancing the portfolio to match strategy

With the opportunity to invest in different assets with different working interests, it is possible to create many combinations that make up the portfolio, and the benefit of risk reduction compared to investing in a single asset has been shown in Section 10.3.4. This leaves the question as to what portfolio choice to make. This depends on the risk-reward balance that the investor is seeking, which should be a clear part of the stated corporate strategy. The balance can be viewed from two angles. It is achieved either when the investor maximizes the portfolio value for an accepted level of risk, or when the level of risk is minimized while achieving the investor's target portfolio value. Either way, the portfolio choice should place the investor somewhere on the efficient frontier on a risk-reward plot, as introduced briefly in Section 8.6.5, and sketched as a generic plot in Fig. 10.7.

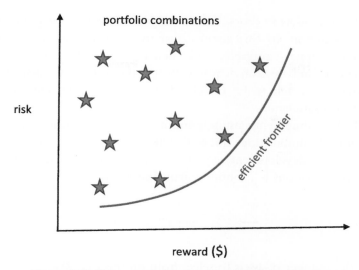

Figure 10.7 Risk-reward plot showing the efficient frontier.

Decision-making with reference to a strategy is key; without a strategy, it becomes difficult to optimize a portfolio, since there is no clear target. Setting a target in terms of rate of return will assist in screening individual projects but not in optimizing a portfolio, since the portfolio IRR is not the sum of the individual asset IRRs. A target level of risk is also challenging to prescribe in a quantitative sense, since it is a ratio, if defined by σ/μ, as in this chapter. Risk can only realistically be described qualitatively as high, medium or low risk. Setting the strategy in terms of a given reward such as hydrocarbon reserves or overall monetary value (e.g. NPV) is a more quantitative and understandable target. In this case, the balanced portfolio should achieve its target at the minimum level of risk. The following example will take the approach of setting value as the strategic target for the portfolio.

Returning to the assets introduced in Table 10.1 as opportunities for investment, suppose that the corporate strategy is to create a value of $2bn by developing a selection of these fields. The volumetric ranges in Table 10.1 have been used, along with FDPs to generate an NPV range for each asset by the method described in Section 10.3.

With seven assets to choose from, many combinations are possible, and we shall assume in the first instance that there are no constraints on the investment capital available, no demands on the sequence of development, and no correlation between the values of each asset. If asset values were felt to be correlated this could be introduced in the Monte Carlo simulation.

With these simplifying assumptions, it is possible to create the risk and reward for all combinations. Note that the permutations would only be required if the development sequence were important. The number of possible combinations is a routine calculation given by

$$\sum_{r=1}^{r=n} {}^nC_r = \sum_{r=1}^{r=n} \frac{n!}{r!*(n-r)!}$$

where

n = the number of assets to select from (in this case 7)

r = the number of assets selected (in this case any number from 1 to 7)

which in the case of this example is 127 combinations. This is not a realistic number of combinations to model in Excel, and this is why more sophisticated algorithms such as the modelling software in Refs. [3–5], or a great deal of patience, are helpful.

Modelling is performed for a selection of portfolio choices in Excel using @RISK as illustrated in Table 10.6. The range of NPVs for each asset is represented by a log-normal input, and combinations of assets are added together probabilistically in @RISK to generate an NPV distribution for the portfolio, whose mean and standard deviation are then used to calculate the risk. Each portfolio choice is then characterised by a reward (mean NPV) and a risk (s.d. NPV/mean NPV). The number of portfolios investigated was filtered by neglecting combinations that are not close to achieving the $2bn target set out in the strategy.

Table 10.6 Excel model to create portfolio NPV reward and risk using @RISK.

Asset	Short name	Input	Value ($m)		Risk
			Mean	s.d.	
Radon	R	= RiskLognorm(F5,G5,RiskShift(0),RiskStatic(F5))	200	80	0.4
Strontium	S	= RiskLognorm(F6,G6,RiskShift(0),RiskStatic(F6))	650	240	0.37
Titanium	T	= RiskLognorm(F7,G7,RiskShift(0),RiskStatic(F7))	120	20	0.17
Uranium	U	= RiskLognorm(F8,G8,RiskShift(0),RiskStatic(F8))	420	120	0.29
Vanadium	V	= RiskLognorm(F9,G9,RiskShift(0),RiskStatic(F9))	1100	500	0.45
Wolfram	W	= RiskLognorm(F10,G10,RiskShift(0),RiskStatic(F10))	500	180	0.36
Xenon	X	= RiskLognorm(Fll,Gll,RiskShift(0),RiskStatic(Fll))	850	400	0.47
	V + X	= RiskOutput() + SUM(E9 + E11)	1950	644	0.33
	X + W + U	= RiskOutput() + SUM(E11 + E10 + E8)	1770	462	0.26
	R + S + T + U + W	= RiskOutput() + SUM(E5 + E6 + E7 + E8 + E10)	1890	350	0.19
Portfolio	V + W + R	= RiskOutput() + SUM(E9 + E10 + E5)	1800	537	0.3
	X + W + V	= RiskOutput() + SUM(E11 + E10 + E9)	2450	667	0.27
	U + W + V	= RiskOutput() + SUM(E8+E10 + E9)	2020	548	0.27
	U + W	= RiskOutput() + SUM(E8 + E10)	920	233	0.25
	U + V	= RiskOutput() + SUM(E8 + E9)	1520	519	0.34

The results of this modelling are shown in Fig. 10.8 on a risk-reward plot.

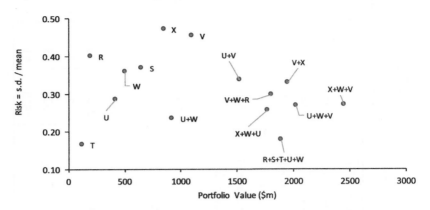

Figure 10.8 Risk-reward plot for single assets and portfolio options.

The efficient frontier curve can now be sketched onto the risk-reward plot, as shown in Fig. 10.9. The curve is the south-east boundary of all combinations modelled. The ideal position for an investment would be in the south-east corner of the plot, high reward with low risk. However, with the portfolio options investigated, the closest that any portfolio can come to this is by being somewhere on the efficient frontier.

If the company strategy is to achieve a target of $2bn, this can be achieved by selecting the portfolio comprising assets (U,W,V).

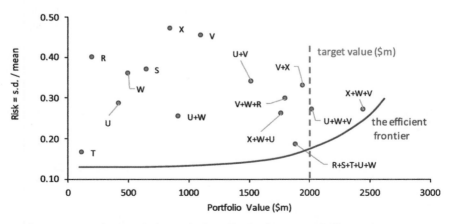

Figure 10.9 Risk-reward plot with the efficient frontier and the strategic target.

Remember that not all combinations have been investigated here. There were 127 possibilities to choose from, and a more thorough investigation may reveal further reasonable alternatives lying close to the efficient frontier.

Nevertheless, from the options investigated, selecting the portfolio of five assets (R,S,T,U,W) almost achieves the target NPV but with considerably less financial risk, while selecting (X,W,V) generates a higher portfolio value for a very similar level of risk as (U,W,V). The investigation has demonstrated that combinations such as (V,X) or (U,V) or (W,V,R) lie behind the efficient frontier, and should be screened out, leaving fewer surviving combinations for further consideration.

More qualitative analysis of the surviving options will be required to integrate issues such as

- resource availability to manage five rather than three projects (work force, service providers, fabrication yards, rig availability)
- added value from operating five assets as opposed to three, such as optionality of tying other prospects into a more widespread infrastructure
- benefits of sharing services between facilities, which will be influenced by the relative locations of the assets
- constraints on financial resources, and the implications for phasing of expenditure on the projects
- licence constraints in terms of production periods if the contract is running in a PSC environment

Imposing quantitative constraints such as capital availability on the options may eliminate some combinations, which will change the shape of the efficient frontier. Fig. 10.10 illustrates the efficient frontier moving as capital constraints tighten, eliminating options that fail to meet the commercial constraints, even though they may be technically feasible. Arguably, these constraints should be applied before spending time investigating the combinations that later fail to meet them. However, it is often the case that the portfolio has to be evaluated to understand how close it comes to the commercial constraints, such as maximum exposure, payback or scheduling issues.

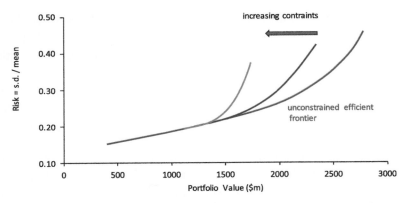

Figure 10.10 Impact of capital constraints on the shape of the efficient frontier.

Generating the efficient frontier determines which portfolio choices are sub-optimal, leaving the remainder of combinations lying on the curve. This still does not solve the question of where on the curve the investor should choose to be. Ultimately, the choice depends on the investor's attitude to the risk-reward balance. A clear strategy is a strong guide to making that decision. If the strategic target is set in terms of achieving a given level of value, or of say reserves additions, then the efficient frontier will guide the investor to the option that delivers this target for the minimum level of risk. Other factors will inevitably come into play as senior management applies less quantitative consideration, but the foundation of the technical and commercial analysis has been built in a quantifiable and auditable manner.

10.5 Gambler's ruin and exploration portfolios

One check that the investor should apply in portfolio analysis is to avoid the risk of what is known as gambler's ruin. This occurs when a run of bad luck creates a series of losses that exceeds the capital available, and the investor goes broke and is unable to continue to participate and earn the long-term EMV.

Investing in a portfolio is a method of reducing the risk of suffering gambler's ruin, by spreading the available capital across a number of assets. If capital is very limited, this can be achieved by taking a smaller working interest in multiple projects. The 127 combinations of the seven opportunities available in Section 10.4 assumed 100% working interest in each one. With the option to dilute the working interest in each asset, many more combinations could be created.

Arps and Arps (Ref. [9]) expressed gambler's ruin in the context of the E&P business, and the following approach to estimating the probability of gambler's ruin from a portfolio of projects is often applied to the exploration setting.

$$v = (1-p)^{C/(w \cdot x)} \qquad (10.5)$$

where
 v = probability of gambler's ruin (fraction)
 p = probability of project success (fraction)
 C = total capital available ($)
 x = capital cost of each project ($)
 w = working interest in each project (fraction)

Arps' approach assumes that the capital cost of each project is identical and that the working interest taken in each project is the same. This is a simplification of the combinations that could be made with multiple project opportunities, but is illustrative of the reduction in the probability of gambler's ruin as the total capital available is spread across a number of projects, as the following example will show.

In the development context, the total capital cost of each project can be considered as its maximum exposure. If the investor has $500 m to invest, and takes 100% working interest in a single project (all eggs in one basket) with a maximum exposure of $500 m and a probability of success of 80%, then the probability of gambler's ruin is 20%, by intuition, or as in the formula

$$v = (1-0.8)^{500/(1.0 \times 500)} = 0.2 \text{ or } 20\%$$

However, if the $500 m available capital is spread between two identical uncorrelated projects by taking only a 50% working interest in each, the probability of gambler's ruin is reduced to

$$v = (1-0.8)^{500/(0.5 \times 500)} = 0.04 \text{ or } 4\%$$

Many examples of this approach and its application to deciding the appropriate working interest to avoid gambler's ruin and to balance a portfolio are described in Ref. [10]. The working interest in ventures can be combined with the investor's attitude to risk, using a concept known as the risk adjusted value (RAV). This is described in detail by Lerche and MacKay (Ref. [11]), with emphasis on exploration portfolio risk. Further examples by M. Walls can be found in Ref. [12].

Note again, that many references define risk as the standard deviation of the values rather than s.d./mean as used in this and other chapters of this book. The benefit of the ratio is that by relating the s.d. to the mean it is dimensionless, and allows asset risk to be compared using the spread of values relative to the mean.

As a final example of attitude to exploration risk and the tension that can exist between the corporate utility and the individual manager's utility, consider the following decision.

An exploration manager has two alternative offshore areas in which to spend the allocated annual $20 m exploration budget. One option is a deep-water prospect called Blue Whale, with a mean success volume (MSV) of 100 MMb reserves and a probability of commercial success (P_c) of 20%. In other words, Blue Whale is a high-risk, high-reward prospect, consuming the whole exploration budget.

The alternative option is to drill five geologically independent prospects in the shallow waters of the more mature shelf area, in prospects known as the Minnows. Each Minnow has a P_c of 40% and an MSV of 20 MMb of reserves, and the drilling cost is $4m per well, so the budget allows all five to be drilled. The options are summarised in Fig. 10.11.

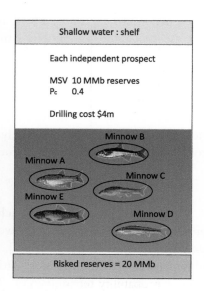

Figure 10.11 Minnows vs Blue Whale exploration opportunities.

The exploration manager is provided with the $20 m exploration drilling budget, but also tasked with adding 25 MMb of reserves to the company, and their next promotion may depend on delivery. Which choice should be made?

The risked reserves in each option is 20 MMb, which is already a dilemma, since neither achieves the target set for the manager. The probability of meeting or exceeding the target when drilling the Blue Whale prospect is the P_c, 20%. Intuitively, drilling the portfolio of the Minnows is a better risk-reward option, and this is correct, as the following analysis will confirm. Table 10.7 lists the possible outcomes when drilling all five Minnow prospects, along with the MSV (refer back to Section 4.2.3) and the probability of occurrence of each outcome.

Table 10.7 Probability and range of reserves from drilling the Minnow prospects.

Number of successes	Mean success volume (MMb)	Probability	Binomial weighting	Weighted probability	Cumulative probability of exceedance
0	0	0.6^5 = 0.078	1	0.08	1.00
1	10	$0.6^4 \times 0.4^1$ = 0.052	5	0.26	0.92
2	20	$0.6^3 \times 0.4^2$ = 0.035	10	0.35	0.66
3	30	$0.6^2 \times 0.4^3$ = 0.023	10	0.23	0.32
4	40	$0.6^1 \times 0.4^4$ = 0.015	5	0.08	0.09
5	50	0.4^5 = 0.010	1	0.09	0.01
				Sum 1.00	

The Minnow prospects are assumed geologically independent, i.e. non-correlated in terms of probability of success. The outcome of one prospect is not influenced by the result of another prospect, otherwise the sequence and dependency would need to be incorporated using Bayes' revision and posterior probability techniques introduced in Section 5.3.1.

With five prospects, there are six possible outcomes for MSVs. The probability of each MSV is multiplied by the number of combinations that can make up that outcome by using the binomial weighting. For example, there are 10 ways that two prospects find oil (AB, AC, AD, AE, BC, BD, BE, CD, CE, DE). A probability tree (see Section 8.2) or a binomial series can also be used here. The weighted probability of all outcomes must sum to 1.0. The weighted probability of each outcome can be accumulated to yield the cumulative probability of exceedance (≥) as plotted in Fig. 10.12.

The MSVs are worked out in discrete steps, so the probability of finding the exploration manager's 25 MMb target is that of finding 30 MMb, which is 32%. Taking a more pragmatic approach and smoothing the results suggest that the probability of finding at least 25 MMb is around 50%. The discrete probability approach is more correct given the single MSVs used in the analysis. The smoothed line would be more justifiable if a probabilistic approach had been taken by combining volumetric input distributions using the Monte Carlo simulation approach. Either way, the probability of meeting the target is better than that of drilling for the Blue Whale, whose probability of exceeding 25 MMb is 20%.

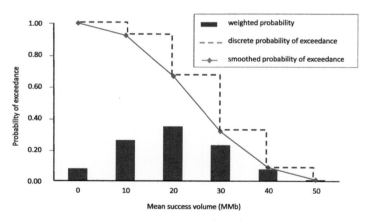

Figure 10.12 Reverse cumulative probability plot for drilling all five Minnow prospects.

The solution to this challenge may feel intuitive; placing all one's eggs in one basket on the high risk, high reward Blue Whale prospect feels less comfortable to the exploration manager (and probably to you) than diversifying the investment into a portfolio of low risk, low reward Minnows, and the statistics bear this out.

However, this choice may not meet the strategy of the company. At corporate level, it may wish to take a different approach to risk by including some high-risk, high-reward prospects in the overall exploration activity. When choices are left to individual managers, hoping to satisfy their own utility (in this case securing a promotion gives great satisfaction), a natural tendency emerges to make decisions based on personal utility, rather than applying the corporate utility.

There are ways to compensate for this human bias. One method is to alleviate the manager of the target that has incentivised the risk-averse behaviour, and instead provide the budget with no specified target. This freedom is liberating, and encourages the exploration manager to become more risk-seeking. This approach has been taken by some companies who appoint a deep-water exploration manager with a brief to be risk-seeking, and issue a different brief to a different shallow water exploration manager.

Another, perhaps more common method is to elevate the decision-making to the corporate board of directors. When presented with an array of options, the board can create a portfolio that matches the corporate risk-reward balance, and it has a view on the corporate utility function, of which the exploration manager may be unaware.

Other quantifiable factors will influence the decision-making, such as the EMV of each development, the technical complexity of a deep-water development, the capex required for development, timing, work force constraints, rig availability and perhaps a non-quantifiable passion to make a significant deep-water discovery. The last point may even be an over-riding consideration at board level. Nevertheless, sound technical and commercial analysis of each investment opportunity is a pre-requisite to good decision-making.

To quote Peter Bernstein (Ref. [13]) 'Without numbers, risk is wholly a matter of gut'.

I hope that this book has provided methods of quantifying value and risk, and demonstrated how to make investment decisions based on structured and auditable analysis rather than instinct; however, spontaneous and convenient instinct may feel.

References

[1] J.L. Jensen, L.W. Lake, P.W.M. Corbett, D.J. Goggin, Statistics for Petroleum Engineers and Geoscientists, Elsevier, 2007, ISBN 978-0-444-50552-1.
[2] D. Freedman, R. Pisani, R. Purves, Statistics, Norton, 2007, ISBN 978-0-393-93043-6.

[3] A. Wilkinson, R. Kupers, Living in the Futures, Harvard Business Review, May 2013. https://hbr.org/2013/05/living-in-the-futures.
[4] P. Wack, Scenarios: uncharted waters ahead, Harv. Bus. Rev. (Sep-Oct 1985) 73–89.
[5] Indeva Ltd, Effective Valuation and Risk Management. http://www.indeva.com/valuation/.
[6] Schlumberger, Merak PEEP Software. https://www.software.slb.com/products/merak.
[7] Aucerna, Aucerna Portfolio Software. https://aucerna.com/products/aucerna-portfolio/.
[8] H.M. Markowitz, Portfolio Selection: Efficient Diversification in Investments, Yale University Press, 1959.
[9] J.J. Arps, J.L. Arps, Prudent risk-taking, J. Pet. Technol. Soc. Pet. Eng. 26 (July 1974) 711–716.
[10] N.A. Mian, Project Economics and Decision Analysis, vol. II, Probabilistic Models, PennWell, 2002, 0-878814-855-8.
[11] I. Lerche, J.A. MacKay, Economic Risk in Hydrocarbon Exploration, Academic Press, 1990, ISBN 978-0-12-444165-1.
[12] M.R. Walls, Combining decision analysis and portfolio management to improve project selection in the exploration and production firm, J. Petrol. Sci. Eng. 44 (2004) 55–65. Elsevier.
[13] P.L. Bernstein, Against the Gods; the Remarkable Story of Risk, Wiley, 1996, 0-471-12104-5.

Abbreviations, symbols and units

Abbreviation	Meaning
API	American Petroleum Institute
BYC	Base Year Cost
CAPEX	Capital expenditure
CAPM	Capital asset pricing model
CCS	Carbon capture and storage
CEO	Chief Executive Officer
CF	Cash flow
CoC	Chance of commerciality
COM	Cost oil maximum
COP	Cessation of Production
CPR	Competent Person's Report
CT	Corporation Tax
CV	Coefficient of variation
D&C	Drilling and completion
DCF	Discounted cash flow
DCQ	Daily Contract Quantity
DPI	Discounted Profitability Index
DTA	Decision tree analysis
DTPI	Discounted Total Profitability Index
E&P	Exploration and production
EBITDA	Earnings before Interest Tax and Depreciation and Amortization
EIU	Economist Intelligence Unit
EL	Economic limit
ELT	Economic limit test
EMV	Expected monetary value
EOR	Enhanced oil recovery
FDP	Field development plan
FEED	Front end engineering design
FID	Final Investment Decision
FLNG	Floating liquefied natural gas
FPSO	Floating production storage and offloading
GAAP	Generally Accepted Accounting Principles
GDP	Gross domestic product
GOR	Gas:oil ratio
GRI	General rate of inflation
GSA	Gas sales agreement
HSE	Health, safety and environment

Abbreviation	Meaning
HSSE	Health, safety, security and environment
IEA	International Energy Agency
IFRS	International Financial Reporting Standards
IM	Information Memorandum
IOC	International oil company
IRR	Internal rate of return
JOA	Joint operating agreement
JV	Joint venture
KPI	Key performance indicator
LCB	Loss carry back
LNG	Liquefied natural gas
LoC	Letter of credit
LPG	Liquefied petroleum gas
MCFS	Minimum commercial field size
MEFS	Minimum Economic Field Size
MOD	Money of the day
MSV	Mean success volume
NASA	National Aeronautics and Space Administration
NBP	National Balancing Point
NFA	No further activity
NGL	Natural gas liquid
NOC	National oil company
NPD	Norwegian Petroleum Directorate
NPV	Net present value
NTG	Net to gross
NYSE	New York Stock Exchange
OPEC	Organization of the Petroleum Exporting Countries
Opex	Operating expenditure
P&A	Plug and abandon
P_c	Probability of commercial success
PCG	Parent company guarantee
P_e	Probability of economic success
P_g	Probability of geological success
PI	Profitability Index
PIR	Profit to investment ratio
PRMS	Petroleum Resources Management System
PSA	Production sharing agreement
PSC	Production sharing contract
PV	Present Value
PVUTC	Present Value Unit Technical Cost

Abbreviation	Meaning
PWRI	Produced water reinjection
QRA	Quantitative risk assessment
R/P	Reserves-to-production ratio
RBL	Reserves-based lending
RDI	Rushmore Drilling Index
RF	Recovery factor
RoW	Rest of World
RR	Reserves replacement ratio
RROR	Real rate of return
RSD	Relative standard deviation
RT	Real terms
SA	Service Agreement
SC	Service Contract (fiscal system context)
SC	Supplementary Charge (UK taxation context)
SEC	Securities and Exchange Commission
SOE	Sub-optimal expenditure
SPD	Special Petroleum Duty
SSV	Subsurface safety valve
TAR	Turn around
TL	Technical limit
TLP	Tension leg platform
TTF	Title Transfer Facility
UCCI	Upstream Capital Costs Index
UKCS	United Kingdom Continental Shelf
UN	United Nations
UOCI	Upstream Operating Costs Index
UTC	Unit technical cost
VOI	Value of information
VPP	Volumetric production payment
WACC	Weighted Average Cost of Capital
WI	Working interest
WTI	West Texas Intermediate
WTO	World Trade Organisation

Symbol	Meaning
σ	Standard deviation
μ	Mean
$\varnothing$	Porosity
S_o	Oil saturation

Unit	Meaning
bbl	Barrel
Bcm	Billion cubic metres
boe	Barrel of oil equivalent
Btu	British thermal unit
J	Joule
Mb/d	Thousand barrels per day
Mbd	Thousand barrels per day
MMb/d	Million barrels per day
MMm3/d	Million standard cubic metres per day
MMscf/d	Million standard cubic feet per day
scm	Standard cubic metre
Tcf	Trillion cubic feet
toe	Tonne of oil equivalent
Tscf	Trillion standard cubic feet

Index

'*Note*: Page numbers followed by "f" indicate figures and "t" indicate tables.'

A

A priori probability, 115
A&D. *See* Acquisition and disposal (A&D)
Abandonment, 275—277
　cost, 281
　fund, 306
　rate, 279
Acceleration projects, 277, 279, 280f
ACQ. *See* Annual Contract Quantity (ACQ)
Acquisition and disposal (A&D), 51—55
Acquisition target, 54
Addition of input variables, 80f
Additional recovery projects, 277, 280f
American Petroleum Institute, 143
Anchoring and adjustment, 234
Andrew Field, 297
Annual Contract Quantity (ACQ), 32
Annual fixed opex, 178
Annual net cash flow, 8
Applying investment themes, 242
Appraisal activity, 6
Appraisal impact on project schedule, 126—128
Appraisal planning, 6
　appraising for comfort, 109
　incorporating imperfect information, 114—122
　justifying reservoir appraisal assuming perfect information, 103—113
　reservoir appraisal
　　planning, 122—130
　　and project assessment, 101—102
Assets
　combining assets
　　using Monte Carlo simulation, 318—320
　　parametrically, 315—318
　defensive, 192
　late-life, 302, 304
　value, 228
Association contracts, 174
Atlanta-based Intercontinental Exchange (ICE), 15

B

Balance sheet value, 51, 62—63
Balancing portfolio to match strategy, 324—330
Banking sector risk, 49—50
Base case, 275—276, 281
Baseline 3-D seismic survey, 122
Bayes, Thomas, 114
Bayes' theorem, 115
　for house break-in as Venn diagram, 116f
　probability tree approach to, 117f, 130f
　as Venn diagram for cancer screening, 129f
Bayesian revision of probabilities, 114—119
　medical example of, 128—130
　using PrecisionTree, 118f, 131f
　of prior and posterior probabilities, 120t
　Venn diagram for, 119f
Bell Laboratories, 252
Benchmark
　crudes, 143
　and industry performance on capital costs, 152—153
Best guess, 234
Beta value, 192
Bid formulation, 96—100
Bidding processes, 45—50
Blue Whale, 332—333
　exploration opportunities, 332f
Bonny Light (Nigeria), 143
Book value. *See* Balance sheet value

341

Bow Tie model, 251–256, 251f
  to manage risk events in well design, 255t
  showing controls and contingencies, 254f
BP's Statistical Review of World Energy 2020, 18–20, 27
Break-even analysis, 48
Break-even oil price, 153t, 211–214, 214f
Brent Blend (W Europe), 143
Brent Crude, 143
Broker's fee, 91
Brown field, 275
Buy-back agreement, 172, 172t

C

Capex. *See* Capital expenditure (Capex)
Capital allowance, 146, 157, 161–162, 178, 201
  straight line and declining balance methods, 158t, 160t
  unit of production method, 159t, 161t
Capital asset pricing model (CAPM), 192
Capital efficiency, 208–221
  ratios, 10, 208t, 209
Capital expenditure (Capex), 7, 12, 145–153, 157–158, 179, 200, 209, 250–251
  costs, 250
  estimation, 146–148
  recovery, 184
  trade-off, 250f
CAPM. *See* Capital asset pricing model (CAPM)
Carbon capture and storage (CCS), 40
Cash, 305
Cash flows (CFs), 127. *See also* Project cash flow
  economic indicators from, 207–208, 208t
CCS. *See* Carbon capture and storage (CCS)
Cervantes, Miguel, 309–310
Cessation of Production (CoP), 13, 151, 295–299

CFs. *See* Cash flows (CFs)
Chance of commerciality (CoC), 52–53
Charge, 74
China–Russia East-Route Natural Gas pipeline, 28
Choke model, 278f
CIS. *See* Commonwealth of Independent States (CIS)
Classes and influence in fiscal systems, 154–155
CoC. *See* Chance of commerciality (CoC)
Coefficient of variation (CV), 11, 262, 310
COM. *See* Cost oil maximum (COM)
Combined discount rate, 202–203
Comfort, appraising for, 109
  appraisal impact on reducing sub-optimal expenditure, 113f
  appraising to add hydrocarbon volume, 110–111
  appraising to avoid sub-optimal expenditure, 111–113
  appraising to prove commerciality, 109–110
  relationships between development cost and field size, 111f
  step-out appraisal, 110f
Commerciality, appraising to prove, 109–110
Commodities, 15, 19–20, 30
  supply and demand curves for, 15–20
Commonwealth of Independent States (CIS), 24
Company valuations, 51
Competent Person's Report (CPR), 55
Composite, 281
  case cash flow model, 284t
  case model, 281, 284–286
Compound interest, 188
  effect on investment, 188t
Concept selection, 244–245
  matrix using updated development concepts from threading, 244f
Concessionary systems, 154–164
Condensate, 136
Consumer Price Index (CPI), 197

Contango, 19–20
Contingencies, 254, 254f
 effect on risk profile, 258f
Contingent resources, 52–53, 169–170
Contractor, 164
 net cash flow, 185
Contractor take, 154, 173
 under pure service contract, 171t
 under risk service contract, 173t
Contractual system, 155
Controls, 254, 254f
 effect on risk profile, 258f
CoP. See Cessation of Production (CoP)
Corporate strategy, 325
Corporation taxes (CTs), 153, 163–164, 302–303
Correlation between asset values, 317, 322, 323t
Correlation coefficient (R), 316
Cost of capital, 191–193
Cost oil, 165–166, 183–184
 available for capex recovery, 184
Cost oil maximum (COM), 165–166, 183–184
Cost–benefit analysis, 256
CostOS from Nomitech, 148–149
Country risk, 49–50
Covariance, 315
 between asset values, 316
Covid-19 pandemic, 20
 impact on energy demand, 21–23
 impact on share price of, 262
CPI. See Consumer Price Index (CPI)
CPR. See Competent Person's Report (CPR)
CPS. See Current Policies Scenario (CPS)
Crystal Ball, 82, 318
CTs. See Corporation taxes (CTs)
Cumulative capex, 179
Currency risk, 49–50
Current Policies Scenario (CPS), 35t
CV. See Coefficient of variation (CV)

D

Daily Contract Quantity (DCQ), 32
Data room, 55

DCF. See Discounted cash flow (DCF)
DCQ. See Daily Contract Quantity (DCQ)
Debt, 134–135
 financing, 62–63
 seniority ranking, 62, 62f
Decision tree
 EMV using, 91–94
 hand-drawn, 106f
 supporting EMV calculation, 96f
 used to calculate break-even probability of success, 294f
Decision tree analysis (DTA), 6, 89–91
Decision-making with reference, 325
Declining balance method, 158, 158t, 160t
Decommissioning, 12–13, 295–306. See also Incremental projects
 activity, 299–301, 299t–300t
 cessation of production, 295–299
 costs, 224–225, 284, 301–302
  fiscal treatment, 302–304
  liabilities, 304–306
  in tax and royalty example, 224t
 economic limit, 295–299, 296f
 potential deferral, 296f
Decommissioning security agreements (DSAs), 305
Defensive asset, 192
Degree of concern, 257–258
Delivery Capacity, 32
Demand curve (D_1), 16
Depletion Contracts, 31
Depreciation, 228
Deterministic methods, 77
Devaluation, 49–50
Development concepts, 232, 240–241
 development options
  for disposal and export, 241t
  for facilities, 240t
  matrix, 242
  for subsurface, 240t
 development planning using scenario approach, 231–233, 232t
 extended test of, 245t
 risk management of, 245–274

Development concepts (*Continued*)
  Bow Tie model, 251–256, 251f
  facilities options for development of an offshore oil field, 263t
  financial adviser's view on risk, 261–268
  risk matrix, 256–259
  risk register, 259
  spider diagrams, 246–251, 248f
  utility, 268–274
  testing alternative development concepts, 244
Discount factor, 189–190
Discounted cash flow (DCF), 3–4, 51
  economic indicators from, 207–208, 208t
Discounted Profit to Total Investment ratio (DPTI), 210
Discounted Profitability Index (DPI), 209
Discounting, 187–196
  compound interest effect on investment, 188t
  cost of capital, 191–193
  internal rate of return, 194–196
  NPV calculation, 189t
  opportunity cost, 193–194
Distribution
  of input variables to estimate UR, 77f
  log-normal, 79
  of proved gas reserves, 27f
  of proved oil reserves, 23f
  of proved reserves, 24
  types used in volumetric estimates, 78f
DPI. *See* Discounted Profitability Index (DPI)
DPTI. *See* Discounted Profit to Total Investment ratio (DPTI)
DSAs. *See* Decommissioning security agreements (DSAs)
DTA. *See* Decision tree analysis (DTA)
Dubai Crude (Middle East), 143

E

E&P investment. *See* Exploration and production investment (E&P investment)
Earnings. *See* Net income

Earnings before interest, taxes, depreciation and amortisation (EBITDA), 51, 216–217
Economic indicators, 221–222
  from cash flow and DSC, 207–208, 208t
  decommissioning cost, 224–225
    in tax and royalty example, 224t
  efficiency indicators
    break-even price, 211–214, 214f
    capital efficiency, 208–221
    cautionary use of Excel = IRR function, 220–221
    interpretation and application of IRR, 214–220
    technical cost efficiency, 210–211
    unit technical cost, 211–214, 212f
  information from, 222t
  net cash flow *vs.* net income, 228–229
  selecting between projects, 225–227
  for tax and royalty example, 223t
Economic lifetime, 180, 279–280, 302
Economic limit (EL), 8, 279, 295–299, 296f
Economic limit test (ELT), 150, 296
Economic rent, 173
Economist Intelligence Unit (EIU), 49–50
ECOPETROL, 174
Edmonton Par (Canada), 143
Effectiveness, 73
Efficiency indicators
  break-even price, 211–214, 214f
  capital efficiency, 208–221
  cautionary use of Excel = IRR function, 220–221
  interpretation and application of IRR, 214–220
  technical cost efficiency, 210–211
  unit technical cost, 211–214, 212f
Efficiency ratio incorporating lifetime costs, 210t
Efficient frontier, 324–330, 325f
EIU. *See* Economist Intelligence Unit (EIU)
EL. *See* Economic limit (EL)
Electrification, 22–23, 26, 36–37, 41
ELT. *See* Economic limit test (ELT)

EMV. *See* Expected monetary value (EMV)
Enhanced oil recovery project (EOR project), 298
Entrepreneurial individuals, 58
EOR project. *See* Enhanced oil recovery project (EOR project)
Equilibrium, 16f, 17
　price, 15, 17
Equity, 134—135
　determination, 58—61
　　basis for, 59t—60t
　financing, 63—64
　share in project, 59
Ernst and Young research (EY research), 152
EU. *See* Expected Utility (EU)
European Union (EU), 26
Excel = IRR function, 220—221
Excel = SUMPRODUCT function, 245
Expected monetary value (EMV), 4, 47—48, 50, 88, 105, 261, 321
　based on MSV development only, 92f
　based on three outcomes that exceed MCFS, 93f
　based on three volumetric outcomes, 93f
　using decision trees, 91—94
Expected Utility (EU), 269
　development decision
　　for risk-averse NewCo using, 269f
　　for risk-seeking MagCo using, 270f
Exploration, 5, 69
　opportunity, 50
　risk in, 70—100, 71f, 85f
　　prospect risking, 72—74
　　risk-reward balance, 88—89, 88f
　summarising opportunity, 94—96, 95t
　uncertainty in, 70—100, 71f
　　bid formulation, 96—100
　　combining exploration risk and volumetric uncertainty, 83—88
　　cumulative probability curves of STOIIP, 84f
　　DTA, 89—91
　　EMV using decision trees, 91—94
　　factors influencing exploration bid value, 99t
　　volumetric uncertainty, 75—83
Exploration and production investment (E&P investment), 1. *See also* Gaining access to E&P opportunity
　cumulative cash flow forecast for oil project with tax and royalty, 9f
　field lifecycle, 2f
EY research. *See* Ernst and Young research (EY research)

F

FDP. *See* Field development plan (FDP)
FEED. *See* Front End Engineering Design (FEED)
FID. *See* Final Investment Decision (FID)
Field development plan (FDP), 231
Field life cycle, 70
Final Investment Decision (FID), 152
Financial adviser's view on risk, 261—268
Financing project, 61—67
　debt financing, 62—63
　debt seniority ranking, 62, 62f
　equity financing, 63—64
　Islamic finance, 66—67
　mezzanine financing, 64—65
　reserves-based lending, 65—66
　volumetric production payments, 66
Fiscal allowances. *See* Tax—deductions
Fiscal costs. *See* Tax—deductions
Fiscal rules, 146
Fiscal systems, 8, 153—176
　association contracts, 174
　classes and influence, 154—155
　petroleum fiscal systems, 155f
　　comparison of, 175—176, 175t
　　around world, 176f
　production sharing contracts, 164—170
　　split of production under, 165f
　service contracts, 170—173

Fiscal systems (*Continued*)
  pure service contracts, 171—172, 171t
  risk service contracts, 172—173, 173t
  tax and royalty systems, 155—164
Fiscal treatment of decommissioning costs, 302—304
Fiscal uncertainties, 49—50, 49t
Fixed opex, 200
Fixed platform, 148
Floating LNG (FLNG), 28—29
Floating production, storage and offloading facility (FPSO), 148
Floating production storage and offloading vessel (FPSO vessel), 240
Forecasts of global energy supply and demand, 33—41
  energy demand forecasts and role for renewable, 34—37
  gas demand forecasts, 39—41
  groups annually issuing, 34t
  IEA scenarios used for forecasting, 35t
  oil demand forecasts, 37—38
  world primary energy demand by fuel and related CO_2 emissions, 35f
Forties Field, 162—163
Forties Field in UK Central North Sea, 61
FPSO. *See* Floating production, storage and offloading facility (FPSO)
FPSO vessel. *See* Floating production storage and offloading vessel (FPSO vessel)
Free gas, 76
Front End Engineering Design (FEED), 147
Full-year discounting, 190, 190f
Future sum of money (C_t), 188—189

G

GAAP. *See* Generally Accepted Accounting Principles (GAAP)
Gaining access to E&P opportunity, 4. *See also* Exploration and production investment (E&P investment)
  acquisition and disposal, 51—55
  bidding processes, 45—50
  equity determination, 58—61
    basis for, 59t—60t
  financing project, 61—67
  joint venture and partnership, 57—58
  merger, 55—56
  options for, 43—44, 44f
  strategic choices for investment, 44—45
  technical and fiscal uncertainties, 49—50, 49t
  unitisation, 58—61
Gambler's ruin and exploration portfolios, 330—335
Gas, 15, 26, 144
  consultants, 301—302
  demand forecasts, 39—41
  prices, 320
    assumptions, 144—145
  sales agreements and influence on gas sales price, 31—33
Gas formation volume factor (B_g), 76
Gas sales agreement (GSA), 31
Gas:oil ratio (GOR), 279
GDP. *See* Gross domestic product (GDP)
Gearing, 134—135
General rate of inflation (GRI), 142
Generally Accepted Accounting Principles (GAAP), 229
Genesis Oil, 301—302
Geostatistical methods, 77
Global distribution of proved gas reserves, 27, 27f
Global energy supply and demand forecasts of, 33—41
  energy demand forecasts and role for renewable, 34—37
  gas demand forecasts, 39—41
  groups annually issuing, 34t
  IEA scenarios used for forecasting, 35t
  oil demand forecasts, 37—38
  world primary energy demand by fuel and related CO_2 emissions, 35f
  historical trends, 20—33

components of long-term gas sales agreement, 31f
distribution of proved gas reserves, 27f
distribution of proved oil reserves, 23f
gas sales agreements and influence on gas sales price, 31—33
historical gas supply, 26—30
historical oil supply, 23—26
immediate impact of Covid-19 on energy demand, 21—23
supply and demand curves for general commodities, 15—20
Global power generation, 36—37, 36f
GOM. *See* US Gulf of Mexico (GOM)
GOR. *See* Gas:oil ratio (GOR)
Government take, 173, 209
under pure service contract, 171t
under risk service contract, 173t
Green curve, 261—262
GRI. *See* General rate of inflation (GRI)
Gross domestic product (GDP), 22—23
Gross revenue, 179
Gross Split PSC, 170
Ground truth, 105, 113, 115, 119
GSA. *See* Gas sales agreement (GSA)
Guidance, 300—301

H

Hand-drawn decision tree, 106f
Handy statistical rules, 80
HCIIP. *See* Hydrocarbon initially in place (HCIIP)
Health, safety or environmental (HSE), 151, 154
Hedging, 317—318, 320
Henry Hub, 145
Historical gas supply, 26—30
Historical oil supply, 23—26
Host government take, 153—176
HSE. *See* Health, safety or environmental (HSE)
Hurdle rate of return, 215
Hybrid systems, 176
Hydrocarbon initially in place (HCIIP), 76, 314

Hydrocarbon volume, appraising to add, 110—111
Hydrocarbons, 75

I

ICE. *See* Atlanta-based Intercontinental Exchange (ICE)
IEA. *See* International Energy Agency (IEA)
IFRS. *See* International Financial Reporting Standards (IFRS)
IHS Markit, 97, 301
IMO. *See* International Maritime Organisation (IMO)
Imperfect information, 114—122
Bayesian revision of probabilities, 114—118
from seismic survey, 118—122
updated DTA with imperfect information, 121f
Incremental projects. *See also* Decommissioning
acceleration *vs.* additional recovery, 280f
activities, 276t—277t
bottlenecks identification, 278f
composite case cash flow model, 284t
current well completion, 293f
economics, 281—287
identification, 275—280
incremental net cash flow for scale inhibitor injection project, 294f
incremental production from infill well, 282f
isolating incremental net cash flow, 282f
opportunities
evaluation, 12
through producing lifetime, 276f
ranking, 287—295, 288t
purely incremental cash flow elements analysis, 285t
reference case cash flow model, 283t
Individual bids, 70
Indonesian Government, 164
Inflation into project cash flow, 196—204, 198f, 200t

Inflation into project cash flow (*Continued*)
  UK and US CPI inflation rates (1970−2020), 196f
Input variables
  addition and multiplication of, 80f
  distribution to estimate UR, 77f
  to estimate recoverable hydrocarbon volumes, 75f
Insolvency, 49−50
Internal rate of return (IRR), 10, 194−196, 202t, 214
  conundrum, 219f
  at cost of capital, 215f
  as hurdle rate, 221
  as screening tool with target hurdle rate, 216f
International Energy Agency (IEA), 22, 26, 140
  New Policies Scenario, 140−141
International Financial Reporting Standards (IFRS), 229
International Maritime Organisation (IMO), 300
International oil companies (IOCs), 2, 58, 164, 168, 174
International Petroleum Exchange (IPE), 15, 141
Interpretation and application of IRR, 214−220
IOCs. *See* International oil companies (IOCs)
IPE. *See* International Petroleum Exchange (IPE)
IRR. *See* Internal rate of return (IRR)
Islamic finance, 66−67
Islamic Law. *See* Sharia principles
Isthmus (Mexico), 143

J
JOA. *See* Joint Operating Agreement (JOA)
Johan Svedrup Field, 2
Joint Operating Agreement (JOA), 57−58, 60−61
Joint ventures (JVs), 4
  arrangements, 43−44
  and partnership, 57−58
Junior debt. *See* Subordinated debt
JVs. *See* Joint ventures (JVs)

K
Key performance indicator (KPI), 25−26
Keynesian economics, 154
KPI. *See* Key performance indicator (KPI)

L
Late-life asset, 302, 304
Late-life production, 295
LCB. *See* Loss carry back (LCB)
Letter of Credit (LoC), 305
Licence to operate, 277, 286−287, 288t, 289
Licensing rounds, 45, 45f, 47
Lifting rights, 58
Liquefied natural gas (LNG), 28, 39
  global gas prices, 30f
  imports by source, 29f
  Middle East growth in, 39
  plant construction project, 219
  transportation of, 29
LNG. *See* Liquefied natural gas (LNG)
LoC. *See* Letter of Credit (LoC)
Log-normal distribution, 79
  values, 79f
Longer-term post-Covid-19 recovery of demand for energy, 22
Loss carry back (LCB), 302−304
Lost time incidents (LTIs), 257
Low-cost incremental projects, 286

M
MagCo, 269, 270f
Marker crudes, 143
Market capitalization, 193
Markowitz diversification, 13, 321−324
Marshal, Alfred, 15−16
Materiality, 220
Mature field, 275, 286
Maximum exposure, 9, 225, 227, 286
MCApp shareware, 82

MCFS. *See* Minimum commercial field size (MCFS)
Mean, 79, 315
Mean success volume (MSV), 87, 332, 334
MEFS. *See* Minimum Economic Field Size (MEFS)
Mergers, 4, 55–56
Mezzanine financing, 64–65
Mid-year discounting, 190, 190f
  NPV calculation using, 191t
Middle East R/P, 27
Minimum commercial field size (MCFS), 86, 86f, 109–110
  appraisal to prove MCFS, 110f
  deterministic outcomes from exploration cases, 87f
Minimum Economic Field Size (MEFS), 86
Minimum nomination, 32
Minimum take or pay quantity, 33
Minnows, 332–333
  exploration opportunities, 332f
  probability and range of reserves from drilling, 333t
MOD. *See* Money of the day (MOD)
Monetary reward, 5
Money of the day (MOD), 141–142, 199, 201, 224
  oil price assumptions, 142f
Monte Carlo approach, 87
Monte Carlo simulation, 81–82, 81f, 122–123, 235, 311, 313–314, 334
  combining assets using, 318–320
MSV. *See* Mean success volume (MSV)
Multi-disciplinary effort, 98
Multiple deterministic modeling, 235
Multiplication of input variables, 80f

N

National Balancing Point (NBP), 145
National oil companies (NOCs), 2, 57–58, 155, 306
Natural diversification, 320

Natural gas liquids (NGLs), 18–19, 136
Natural Gas price indices in year 2020, 144t
NBP. *See* National Balancing Point (NBP)
Negative correlation coefficients, 313, 317
Net cash flow
  net income, *vs.*, 228–229
  of phased project, 220f
Net income, 228
Net payoffs, 272
Net present value (NPV), 4, 10, 10t, 65, 88, 96–97, 105, 127t, 189, 209, 264–265, 286
  calculation, 65
    using mid-year discounting, 191t
    at various discount rates, 194t
    at cost of capital, 215f
Net profit oil to contractor, 185
Net-zero emissions, 40
Netback price, 151–152
New York Mercantile Exchange (NYMEX), 145
New York Stock Exchange (NYSE), 51
NewCo, 269, 269f
Newcomer, 162–163
NFA. *See* No further activity (NFA)
NGLs. *See* Natural gas liquids (NGLs)
No further activity (NFA), 12, 275–276
NOCs. *See* National oil companies (NOCs)
Nominal terms money. *See* Money of the day (MOD)
Norwegian Petroleum Directorate (NPD), 45–47
  offshore licensing awards from, 46f
Norwegian taxation system, 164
NPD. *See* Norwegian Petroleum Directorate (NPD)
NPV. *See* Net present value (NPV)
NYMEX. *See* New York Mercantile Exchange (NYMEX)
NYSE. *See* New York Stock Exchange (NYSE)

O

Office of National Statistics (ONS), 196–197
Offshore discovery, 104
Offshore licensing awards from NPD, 46f
Offshore Petroleum Regulator for Environment and Decommissioning (OPRED), 300–301
OGA. *See* UK Oil and Gas Authority (OGA)
Oil demand, 140
  in Asia Pacific, 38
  forecasts, 37–38
Oil formation volume factor (B_o), 76
Oil price, 178, 320
  assumptions, 139–144
    history of oil price over last century, 139f
  forecast, 143
Oil-water contact depth (OWC), 236
Omani Government, 57
ONS. *See* Office of National Statistics (ONS)
OPEC Reference Basket, 143
OPEC+. *See* Organization of Petroleum Exporting Countries plus Russia (OPEC+)
Operating expenditure (Opex), 8, 145–153, 157–158, 179, 209, 250–251
  estimation, 148–152
  revenue and opex trends in late field life, 151f
Opex. *See* Operating expenditure (Opex)
Opportunity
  cost, 193–194
  evaluating field development, 7–10
  exploration, 94–96, 95t
  incremental projects
    evaluation, 12
    through producing lifetime, 276f
  ranking, 287–295, 288t
OPRED. *See* Offshore Petroleum Regulator for Environment and Decommissioning (OPRED)
Option value, 105
Organization of Petroleum Exporting Countries plus Russia (OPEC+), 21–22
Original book value, 228
OSPAR Convention, 300
Out-of-round bids, 47
Outcome of contract extension discussions, 169–170
OWC. *See* Oil-water contact depth (OWC)
Ownership of production, 155–156, 164

P

P&A. *See* Plug and abandonment (P&A)
P10, 88, 105–106, 139, 312–313
P50, 88, 104–106, 112, 139, 312–313
P90, 83, 88, 106, 119, 139, 312–313
Parametric method, 80
Parent company guarantee (PCG), 305
Paris Agreement on Climate Change, 2–3, 26
Passive investment in E&P sector, 43
Pay-as-you-go time, 8, 286
Payback projects, 226
Payback time, 286
PCG. *See* Parent company guarantee (PCG)
PD. *See* Proved developed (PD)
PDNP. *See* Proved Developed Non-Producing (PDNP)
PDO. *See* Petroleum Development Oman (PDO)
PDP. *See* Proved Developed Producing (PDP)
Peak Supply Contract, 31
Perfect information, 6, 103–114. *See also* Imperfect information
Petroleum Development Oman (PDO), 57
Petroleum Economist, 45
  upcoming licensing round overview from, 45f
Petroleum fiscal systems, 153, 155f

comparison of, 175–176, 175t
around world, 176f
Petroleum Resources Management
  System (PRMS), 51, 109
  framework for reporting resources, 52f
  prospective resources within, 53
Petroleum Revenue Tax (PRT),
  163–164, 303–304
Petroleum system elements, 72t
Physical commodity, 15–16
PI. *See* Profitability Index (PI)
PIR. *See* Profit-to-Investment Ratio
  (PIR)
Plausible combinations of processing and
  export options, 241f
Play Chance, 73
  risking for P_g, 74t
Plug and abandonment (P&A), 301
Political risk, 49–50
Portfolio effect
  on value and risk, 312–324
  on volumes and risk, 310–312
Portfolio investment, 13–14
Portfolio management
  balancing portfolio to match strategy,
    324–330
  combining assets
    using Monte Carlo simulation,
      318–320
    parametrically, 315–318
  Excel model to create portfolio NPV
    reward and risk, 327t
  gambler's ruin and exploration
    portfolios, 330–335
  Markowitz diversification, 321–324
  portfolio effect
    on value and risk, 312–324
    on volumes and risk, 310–312
  probability and range of reserves from
    drilling Minnow prospects, 333t
  risk reduction through diversification
    options, 324t
  systematic and unsystematic portfolio
    risk, 320–321
  value and risk management, 309–310
Positive correlation coefficients, 313
Positive net trade values, 38
Post-discovery, 104

uncertainty in reservoir extent
  post-discovery well X-1, 104f
Posterior probability, 115
Practicality, 256
Pre-tax cost per barrel analysis, 291
PrecisionTree, 117, 130
  Bayesian revision of probabilities using,
    118f
  probability chart for development
    example, 273f
  version of appraisal VOI, 107, 107f
Preference shareholders, 64
Presence, 73
Present Value (PV), 9–10, 105, 188
Price inelasticity, 18
Price risk, 66
PRMS. *See* Petroleum Resources
  Management System (PRMS)
Probabilistic methods, 77
Probability from drilling Minnow
  prospects, 333t
Probability of commercial success (P_c),
  48, 53, 86, 86f, 88
Probability of economic success (P_e), 86
Probability of exceedance curve. *See*
  Reverse cumulative probability
  density function
Probability of geological success (P_g), 48,
  73
  grouped terms for estimation, 74t
  Play and Prospect Specific risking for,
    74t
Probability tree approach, 236, 237f
  to Bayes' theorem, 117f, 130f
Produced water reinjection (PWRI),
  277
Production profile and sales price,
  136–145
  benchmarks and industry performance
    on capital costs, 152–153
  breakeven oil prices, 153t
  capex, 145–153
  gas price assumptions,
    144–145
  for oil field with associated gas sales,
    137t
  oil price assumptions, 139–144
  opex, 145–153

Production sharing agreement (PSA), 151
Production sharing contracts (PSCs), 8, 164—170
  project cash flow under, 182—186, 183t, 186f
  split of production under, 165f
Profit. See Net income
Profit oil, 184
  to contractor, 184
  to government, 184
Profit-to-Investment Ratio (PIR), 209—210
Profitability Index (PI), 209
Project, 52
  assessment, 101—102
  complexity, 289—290, 289f
  criteria for defining success of, 252t
  equity share in, 59
  financing, 66
  maturity, 52
  net cash flow, 134, 228
  schedule, 250—251
  scope, 134
Project cash flow, 7—10, 176
  components, 133—135, 134t
  data requirement for project cash flow generation, 135t
  discounting, 187—196
  fiscal systems, 153—176
  incorporating inflation into, 196—204, 198f, 200t
  production profile and sales price, 136—145
  project funding and cash generation, 136f
  under PSC terms, 182—186, 183t, 186f
  under tax and royalty terms, 177—182, 177t, 181f
Prospect risking, 72—74
Prospect Specific Chance, 73
  risking for P_g, 74t
Prospective resources, 52—53
Proved developed (PD), 26
Proved Developed Non-Producing (PDNP), 65
Proved Developed Producing (PDP), 65

Proved undeveloped (PUD), 26, 65
PRT. See Petroleum Revenue Tax (PRT)
PSA. See Production sharing agreement (PSA)
PSCs. See Production sharing contracts (PSCs)
PUD. See Proved undeveloped (PUD)
Pure service contracts, 171—172, 171t
Purely incremental cash flow elements analysis, 285t
PV. See Present Value (PV)
PWRI. See Produced water reinjection (PWRI)

Q

Qualitative matrices, 289—290
Quantitative Risk Analysis (QRA), 11, 259
QUE$TOR, 147, 301—302

R

R-value, 272
R/P ratio. See Reserves-to-Production ratio (R/P ratio)
Radon plus Xenon, 311—312
Range of reserves from drilling Minnow prospects, 333t
Range of uncertainty, 52
Ranking incremental projects opportunities, 287—295, 288t
RAV. See Risk adjusted value (RAV)
RBL. See Reserves-based lending (RBL)
RDI. See Rushmore Drilling Index (RDI)
Real rate of return (RROR), 10, 201, 202t, 204, 214
  as hurdle rate, 221
Real terms (RT), 141—142
  oil price assumptions, 142f
Recoverable distributions, 96
Reference, 281
  case
    cash flow model, 283t
    project, 134
  date, 188
Refinitiv, 49—50

Regulations, 300–301
Relative standard deviation (RSD), 262
Replacement ratio (RR), 25–26
Reserves, 52
  categories of reserves under PRMS, 53t
  portfolio evaluation based on classes and categories of, 55t
  reporting guidelines, 51
Reserves-based lending (RBL), 65–66
Reserves-to-Production ratio (R/P ratio), 24, 25f
Reservoir
  parameters influencing recoverable volumes, 238t
  realisation, 232–239
  uncertainty long list, 238–239, 238t, 239f, 252
Reservoir appraisal
  assuming perfect information, 103–113
  planning, 122–130
    impact of appraisal on project schedule, 126–128
    appraisal tools, 123–126
    medical example of Bayesian revision, 128–130
    reservoir appraisal tools and interpreted information, 125t
    target of appraisal, 122–123, 124f
  and project assessment, 101–102
Resources for Future (RFF), 34
Retail Price Index (RPI), 197
Return on average capital employed (ROACE), 216–217, 217f
Revenues, 297
  source, 136–145
Reverse cumulative probability density function, 85
RFCT. See Ring Fence Corporation Tax (RFCT)
RFF. See Resources for Future (RFF)
Ring fence, 161–163
Ring Fence Corporation Tax (RFCT), 162–163, 304
Risk
  analysis, 10–11
  events, 251, 253, 256
    risk matrix with, 257f

in exploration, 70–100, 71f
  prospect risking, 72–74
  risk-reward balance, 88–89, 88f
management, 309–310
matrix, 256–259
  with risk event, 257f
portfolio effect on, 310–324
reduction through diversification options, 324t
register, 259
  with pre-and post-measure assessment, 260f
service contracts, 172–173, 173t
@RISK, 82, 89, 91, 270, 313, 318, 326
  Excel model to create portfolio NPV reward and risk using, 327t
  simple decision tree in, 91f
Risk adjusted value (RAV), 331–332
Risk management of development concepts, 245–274
  Bow Tie model, 251–256, 251f
  development options
    decision tree to address, 264f
    risk and reward table for, 267t
    risk-reward plot for, 267f
    sensitivity analysis on assumption of aquifer performance, 265f
    simplified decision tree with two competing options, 266f
  facilities options for development of offshore oil field, 263t
  financial adviser's view on risk, 261–268
  risk matrix, 256–259
  risk register, 259
  Spider diagrams, 246–251, 248f
  utility, 268–274
Risk tolerance (R), 272, 272f
Risk-reward plot
  with efficient frontier and strategic target, 328f
  showing efficient frontier, 325f
  for single assets and portfolio options, 328f
Risked value, 47–48, 52–54
ROACE. See Return on average capital employed (ROACE)

354 Index

Royal Dutch Shell, 58, 261–262, 318
Royalty, 156–158, 179
  economic model, 246t
  split of revenue under, 157f
  system, 154
Royalty systems, 155–164
  project cash flow under, 177–182, 177t, 181f
    economic indicators, 182t, 187t
RPI. *See* Retail Price Index (RPI)
RR. *See* Replacement ratio (RR)
RROR. *See* Real rate of return (RROR)
RSD. *See* Relative standard deviation (RSD)
RT. *See* Real terms (RT)
Rushmore Drilling Index (RDI), 152

S

SA. *See* Service agreement (SA)
Safety-critical subsurface safety valve (SSV), 292
Sale-and-buy-back contract, 67
Sales price, 136–145
Sanction. *See* Final Investment Decision (FID)
SC. *See* Service contracts (SC); Supplementary Charge (SC)
Scenario, 233
  approach, 231–233, 232t
Scorpion plot, 290–291
Screening
  and ranking, 225
  rate, 215
SDS. *See* Sustainable Development Scenario (SDS)
Securities and Exchange Commission (SEC), 51
Seismic survey, imperfect information from, 118–122
Selecting between projects, 225–227
Semi-submersible production facility, 148
Senior debt, 62–63
Sensitivity analysis on appraisal cost, 108, 108f
Service agreement (SA), 170–171

Service contracts (SC), 170–173
  pure, 171–172, 171t
  risk, 172–173, 173t
Seven Sisters (oil companies), 58
Shares, 64
  issue, 63
Sharia principles, 66
Shell, 57
Shell's Sky scenario, 40
Signature bonus, 47
Simple decision tree, 90, 90f
  in @RISK format, 91f
Simple diversification, 320
Simple sensitivity analysis, 293
Single incremental project, 278
SMART completions, 240
SOE. *See* Sub-optimal expenditure (SOE)
Sonangol-Sinopec in Angola, 97
Sovereign risk, 49–50
SPD. *See* Special Petroleum Duty (SPD)
SPE Petroleum Resource Management System Guidelines, 77
Special Petroleum Duty (SPD), 163–164
Spider diagrams, 246–251, 248f
  input data for, 247t
  use to indicate potential trade-offs, 249f
Sponsor, 63, 66–67
SSV. *See* Safety-critical subsurface safety valve (SSV)
Stage Gate process, 101–103, 102f, 103t, 231
Standard deviation, 79, 315
Stated Policies scenario (STEPS), 35–38
  solar PV sources, 37
Stated Policies Scenario, 26, 35t
Statoil, 166
Step-out appraisal, 110f
STEPS. *See* Stated Policies scenario (STEPS)
Stock tank oil initially in place (STOIIP), 96
  cumulative probability curves of, 84f
STOIIP. *See* Stock tank oil initially in place (STOIIP)

Straight line capital allowance method, 158, 158t, 160t
Stranded gas, 28–29
Strategic choices for investment, 44–45
Strontium (S), 311
Sub-optimal expenditure (SOE), 111, 112f
　calculation due to over-expenditure and under-expenditure, 112
Sub-optimal expenditure, appraising to avoid, 111–113
Subordinated debt, 63
Subsurface uncertainty, 233–239, 253
Supplementary Charge (SC), 163–164, 302–303
Supply and demand curves
　for general commodities, 15–20
　for oil, 18f
　primary energy consumption and sources, 20f
　shifts in, 17f
Supply contracts, 31
Sustainable Development Scenario (SDS), 35t, 36, 140–141
Swanson's approximation, 82
　P90 in, 83
　to weighting discrete values, 82f
Swanson's rule, 105, 312–313
Systematic portfolio risk, 320–321
Systematic risk. See Undiversifiable risk

T

Tapis Crude (Singapore), 143
TAR. See Turn-around (TAR)
Target of appraisal, 122–123, 124f
Tax, 157, 185
　deductions, 157, 179–180
　economic model, 246t
　rate, 178
　split of revenue under, 157f
　systems, 154–164
　　economic indicators, 182t, 187t
　　project cash flow under, 177–182, 177t, 181f
Taxable income, 179–180, 298
Taxation, 157, 229
Teaser discovery, 110–111
Technical costs, 145–153, 172
　efficiency, 210–211
Technical limit (TL), 295

Technical risks, 48
Technical uncertainties, 49–50, 49t
Tension leg platform (TLP), 148
Threading, 242
　across development options matrix, 243f
3-D seismic, 106
Time value of money, 9, 187–196
Title. See Ownership of production
Title Transfer Facility (TTF), 145
TL. See Technical limit (TL)
TLP. See Tension leg platform (TLP)
Tobin, James, 309–310
Tornado diagrams, 122–123
　for volumetric estimate, 123f
Total capex, 179
Total cost oil recovery, 184
Total opex, 178–179
Transportation of gas to markets, 28
Transportation of LNG, 29
Tritium (T), 311–312
Truncated mean. See Mean success volume (MSV)
TTF. See Title Transfer Facility (TTF)
Turn-around (TAR), 278

U

UCCI. See Upstream Capital Costs Index (UCCI)
UK Oil and Gas Authority (OGA), 47, 300–301
UKCS. See United Kingdom Continental Shelf (UKCS)
Ultimate recovery (UR), 77f, 122–123
UN. See United Nations (UN)
Uncertainty
　in exploration, 70–100, 71f
　　bid formulation, 96–100
　　combining exploration risk and volumetric uncertainty, 83–88
　　cumulative probability curves of STOIIP, 84f
　　DTA, 89–91
　　EMV using decision trees, 91–94
　　factors influencing exploration bid value, 99t
　　volumetric uncertainty, 75–83
　in reservoir extent post-discovery well X-1, 104f
　subsurface, 233–239

UNCLoS. *See* United Nations Convention on the Law of Sea of 1982 (UNCLoS)
Undiversifiable risk, 320
Unit of production method, 159, 159t, 161t
Unit technical cost (UTC), 10, 211–214, 212f
United Kingdom Continental Shelf (UKCS), 12–13, 47, 156, 163–164, 305
United Nations (UN), 49–50
United Nations Convention on the Law of Sea of 1982 (UNCLoS), 300
Unitisation, 58–61
Unrecovered capex, 184
Unsystematic portfolio risk, 320–321
Upstream Capital Costs Index (UCCI), 152
Upstream Operating Costs Index (UOCI), 152
UR. *See* Ultimate recovery (UR)
Urals oil (Russia), 143
Uranium (U), 311–312
US Gulf of Mexico (GOM), 301
UTC. *See* Unit technical cost (UTC)
Utility, 268–274
  curve extending to negative outcome, 271f
  typical utility curves, 269f

V

Value
  indicators, 309
  of licence, 70
  management, 309–310
  of oil field assets, 317–318
  portfolio effect on, 312–324
Value of information (VOI), 6, 106, 108–109, 114, 126, 255–256
  PrecisionTree version of appraisal, 107, 107f
  sensitivity to reliability of information, 121, 122f
Vanadium (V), 310–311
Variable opex, 178, 200
  trade-off, 250f

Variance, 79
Venn diagram, 119
  Bayes' theorem as, 129f
  for Bayesian revision of probabilities, 119f
VOI. *See* Value of information (VOI)
Volumes
  appraising to add hydrocarbon volume, 110–111
  portfolio effect on, 310–312
Volumetric production payments (VPP), 66
Volumetric uncertainty, 75–83, 85f
  combining exploration risk and, 83–88
VPP. *See* Volumetric production payments (VPP)

W

Water injection (WI), 264
Weighted average cost of capital method (WACC method), 191–193, 214
WEO. *See* World Energy Outlook (WEO)
WEO 2019 STEPS forecasts, 38
  change in annual gas supply balance by region under, 40f
  geographical change in oil supply and demand under, 37f, 39f
West Texas Intermediate crude oil (WTI crude oil), 21–22, 143
WI. *See* Water injection (WI)
Winner's regret, 50
With/without economic analysis, 281
Wolfram (W), 311–312
Wood Mackenzie's forecasts (WoodMac forecasts), 141, 145
Work programme, 47–48, 50
Working interest, 45, 309, 312, 324, 331
World Energy Outlook (WEO), 34, 140
World Trade Organisation (WTO), 49–50
WTI crude oil. *See* West Texas Intermediate crude oil (WTI crude oil)
WTO. *See* World Trade Organisation (WTO)

CPI Antony Rowe
Eastbourne, UK
February 19, 2021